# OPTICS
## An Introduction for Ophthalmologists

# AN INTRODUCTION FOR OPHTHALMOLOGISTS

# OPTICS

*(Second Edition, Third Printing)*

*By*

**KENNETH N. OGLE, Ph.D.**
**M.D (h.c., Uppsala), D.Sc. (h.c., Colorado College)**

*Head, Section of Biophysics*
*Consultant in Visual Optics to the Section of Ophthalmology*
*Mayo Clinic*
*Professor of Biophysics*
*Mayo Graduate School of Medicine, University of Minnesota*
*Rochester, Minnesota*

**CHARLES C THOMAS • PUBLISHER**
*Springfield • Illinois • U.S.A.*

*Published and Distributed Throughout the World by*
CHARLES C THOMAS • PUBLISHER
BANNERSTONE HOUSE
301-327 East Lawrence Avenue, Springfield, Illinois, U.S.A.

This book is protected by copyright. No part of it may be reproduced in any manner without written permission from the publisher.

© *1961 and 1968, by* CHARLES C THOMAS • PUBLISHER
ISBN 0-398-01417-5
Library of Congress Catalog Card Number: 68-13771

First Edition, 1961
Second Edition, First Printing, 1968
Second Edition, Second Printing, 1971
Second Edition, Third Printing, 1976

*With* THOMAS BOOKS *careful attention is given to all details of manufacturing and design. It is the Publisher's desire to present books that are satisfactory as to their physical qualities and artistic possibilities and appropriate for their particular use.* THOMAS BOOKS *will be true to those laws of quality that assure a good name and good will.*

*Printed in the United States of America*

*To Bettie*

## PREFACE TO THE SECOND EDITION

A TEXT which has been based on a series of lectures may seem entirely satisfactory at the time of writing; but later, when it has been actually used in subsequent classes, its deficiencies seem glaringly evident. The opportunity afforded by Charles C Thomas, Publisher, for a Second and Revised Edition of the book has made possible the correction of many deficiencies. Significant changes have been made, therefore, in many parts of the text. Answers to the exercises have been added, also, in one of the appendices. Again it must be pointed out that this book is to provide only an introduction to the fundamentals of optics. It is hoped that this Second Edition will be even more useful to the student in ophthalmology or the student in any of the visual sciences.

KENNETH N. OGLE

# PREFACE TO FIRST EDITION

IN SPITE of the number of excellent textbooks on optics available today, none seem to fit uniquely the need of the student in ophthalmology. In general, either these textbooks are so detailed that the student loses his way or they contain too much irrelevent material. The present book is based upon a series of lectures prepared for the residents of the Mayo Graduate School of Medicine assigned to the Section of Ophthalmology of the Mayo Clinic. Out of the experience gained by the author from giving these lectures and from discussing their contents with the residents, an attempt is made in this book to present the subject matter in such a way that it meets directly the needs of the student ophthalmologist as preparation for his study of refraction of the eye and as a background for his understanding of physiologic optics and visual physiology.

It is obvious that any textbook written today dealing with the principles of optics can contain little that is new or original. The author makes no pretense of originality. Only the particular selection of topics and the method of presentation justify the writing of the book at all. The author is thus indebted to many authors of books on optics and on visual optics, in particular, from which subject matter has been selected and organized.

Needless to say, this book is just what its title signifies it to be—namely, an introduction to the subject. With this introduction, however, it is hoped that the serious student will find himself equipped to understand more comprehensive and detailed treatises on the subject. Some of these are listed in the Appendix.

The study of optics, as that of any field of science, should start in the laboratory, where the actual phenomena of light and the formation of light images by mirrors, prisms, lenses, and so on, can be observed. In a short monograph such as this one, however, it is expedient to approach the subject by a logical development from statements of the underlying principles of optics, from which one can describe and predict most of the phenomena observable in the laboratory. In studying this book, however, one should always keep in mind, and should try to visualize, the actual physics underlying the discussion at hand. The reader is assumed to be not totally unsophisticated in optical phenomena.

It is virtually impossible to write a useful and authoritative text on optics without the aid of mathematics. In this book the mathematics has been kept as simple as is consistent with rigor, and then the reader is usually prepared for the various steps in derivations. Most medical students will have studied sufficient mathematics to enable them to comprehend easily the algebra used. However, it is hoped that even the casual reader can obtain from this text a knowledge of the basic principles and the essential formulas of geometric optics without specifically following through the various derivations.

K. N. O.

# ACKNOWLEDGMENTS

The author is indebted to many of the residents of the Mayo Graduate School of Medicine in the Section of Ophthalmology of the Mayo Clinic for their suggestions for this monograph. He is especially indebted to Dr. Hugo Bair, but also to Dr. T. G. Martens and Dr. John Dyer, consultants in the Section of Ophthalmology, for their personal assistance, and to Dr. C. Wilbur Rucker, emeritus head of the Section of Ophthalmology, for his encouragement and insistence that the lectures be published.

The monograph would, of course, have been impossible to complete except for the editorial and secretarial work of Miss Lorette Hentges of the Section of Biophysics, the editorial assistance of Dr. Guy Whitehead of the Section of Publications, and the art work by Miss Jane Hanson of the Section of Medical Illustration of the Mayo Clinic.

K. N. O.

# CONTENTS

|  | Page |
|---|---|
| *Preface to the Second Edition* | vii |
| *Preface to the First Edition* | ix |
| *Acknowledgments* | xi |

*Chapter*

I. PHYSICAL CHARACTERISTICS OF LIGHT . . . . . 3
    1. Light . . . . . . . . . . 3
    2. Nature of Light . . . . . . . . 3
    3. Velocity of Light . . . . . . . . 4
    4. Dispersion . . . . . . . . . 5
    5. Physical and Geometric Optics . . . . . 6
    6. Wave Theory of Light . . . . . . . 7
    7. Interference . . . . . . . . . 9
    8. Diffraction Patterns . . . . . . . 13
    9. Diffraction Patterns and the Diffraction Grating . . 13
    10. Polarization of Light . . . . . . . 16
    11. Electromagnetic Theory of Light . . . . . 18
    12. Quantum Optics . . . . . . . . 19
    13. Spectral Energy of Radiation Sources . . . . 19
    14. Interaction of Light with an Object . . . . 21
    15. Absorption . . . . . . . . . 22
   *References* . . . . . . . . . . 25

| Chapter | Page |
|---|---|
| II. GEOMETRIC OPTICS—REFLECTION | 27 |
|    1. Pencils, Rays, Beams of Light | 27 |
|    2. The Concept of Optical Images | 27 |
|    3. Reflection of Light | 29 |
|    4. Image Formation by Reflection | 32 |
|       A. The Plane Mirror | 32 |
|       B. Spherical Mirrors | 34 |
|          1. The Convex Mirror | 36 |
|          2. The Concave Mirror | 41 |
|    5. General Relationships for Mirrors | 43 |
|    *Exercises* | 47 |
| III. REFRACTION OF LIGHT | 49 |
|    1. Refraction of Light | 49 |
|    2. Snell's Law | 49 |
|    3. Critical Angle of Refraction | 53 |
|    4. Image Formation by Refraction at a Plane Surface | 57 |
|    5. Variation of Index of Refraction with Wavelength | 60 |
|    6. Plane Parallels of Glass | 61 |
| IV. OPHTHALMIC PRISMS | 64 |
|    1. Prismatic Deviation | 64 |
|    2. Orientation of Prisms for Minimal Deviation | 67 |
|    3. The Measure of Prismatic Deviation | 69 |
|    4. Effect of Object Distance | 70 |
|    5. Aberrations and Undesirable Properties of Ophthalmic Prisms | 72 |
|    6. Resolution of Prismatic Deviations | 75 |
|    7. The Biprism | 80 |
|    *Exercises* | 81 |
|    *References* | 81 |

| Chapter | Page |
|---|---|
| V. INTRODUCTION TO THE THEORY OF LENSES | 82 |
|     1. Introduction—Basic Concepts and Definitions | 82 |
|         A. Vergence | 82 |
|         B. Lenses | 84 |
|         C. Optic Axis | 85 |
|         D. Focal Points | 85 |
|         E. Focal Plane | 86 |
|         F. Paraxial Rays | 87 |
|         G. Extra-axial Rays | 87 |
|         H. Chief Rays | 87 |
|         I. Power | 88 |
|     2. The Image Formation by an Ideal Infinitely Thin Lens | 88 |
|         A. Graphic Method | 88 |
|         B. Analytic Method | 91 |
|   *Exercises* | 95 |
|   *Reference* | 95 |
| VI. SOME APPLICATIONS OF THIN-LENS THEORY TO VISUAL OPTICS | 96 |
|     1. Angular Magnification | 96 |
|         A. The Magnifying Lens | 97 |
|         B. Subnormal Vision Lenses | 99 |
|         C. Magnification of the Ophthalmoscope | 101 |
|     2. Principles of Ocular Refraction | 102 |
|         A. Emmetropia | 102 |
|         B. Myopia | 104 |
|         C. Hyperopia | 107 |
|         D. Lens Power and Lens Position | 109 |
|         E. Effective Power | 113 |
|   *Exercises* | 114 |
|   *Reference* | 115 |

| Chapter | Page |
|---|---|
| VII. IMAGE FORMATION BY COMBINATIONS OF THIN LENSES | 116 |
|     1. Basic Considerations | 116 |
|     2. Principal Points and Principal Planes | 118 |
|     3. Telescopic Systems | 122 |
|     *Exercises* | 125 |
| VIII. IMAGE FORMATION BY REFRACTION FROM SPHERICAL SURFACES | 127 |
|     1. Refraction by Spherical Surfaces | 127 |
|     2. Magnification of the Image from Single Refracting Surfaces | 132 |
|     3. Combination of two Refractive Surfaces | 134 |
|     4. Thick Lenses | 138 |
|     *Exercises* | 141 |
| IX. OPHTHALMIC OPTICS | 142 |
|     1. Vertex Power of Ophthalmic Lenses | 142 |
|     2. Afocal Lenses | 148 |
|     3. Nodal Points | 149 |
|     4. Location of Nodal Points | 153 |
|     5. The Cardinal Points | 154 |
|     6. The Schematic Eye | 155 |
|     7. Size of Retinal Image | 157 |
|     8. The Reduced Schematic Eye | 158 |
|     *Exercises* | 159 |
|     *References* | 160 |
| X. ASTIGMATISM | 162 |
|     1. Astigmatic Imagery | 162 |

| Chapter | Page |
|---|---|
| 2. Dioptric Components of a Cylindric Lens | 171 |
| 3. Obliquely Crossed Cylinders | 173 |
| 4. The Cross-Cylinder Test Lens | 176 |
| 5. The Maddox Rod | 177 |
| 6. The Scissors Effect of Cylindric Lenses | 177 |
| *Exercises* | 179 |
| *References* | 180 |

XI. ABERRATIONS OF SPHERICAL LENS SYSTEMS . . . . . 181

    1. Spherical Aberration . . . . . . . . 181

    2. Chromatic Aberration . . . . . . . . 182

    3. Coma . . . . . . . . . . . 184

    4. Astigmatism of Oblique Incidence . . . . . 185

        A. The Astigmatism From Tilted Lenses . . . 187

    5. Distortion . . . . . . . . . . 188

    6. Special Topics . . . . . . . . . . 190

        A. Blurred Imagery . . . . . . . . 190

        B. Resolving Power . . . . . . . . 192

        C. Depth of Focus . . . . . . . . 193

    *Exercises* . . . . . . . . . . . 195

    *References* . . . . . . . . . . . 196

XII. CONSIDERATIONS OF OPHTHALMIC LENSES . . . . . 197

    1. Magnification Properties of Ophthalmic Lenses When Used With the Eye . . . . . . . . 197

        A. Ophthalmic Lenses and Eye Movements . . 198

        B. Ophthalmic Lenses and the Size of the Retinal Image . . . . . . . . . . 202

| Chapter | Page |
|---|---|
| 1. Refractive Ametropia | 202 |
| 2. Axial Ametropia | 204 |
| 2. Afocal Lenses | 206 |
| 3. Trial-case Lens Design | 208 |
| 4. Corrected-curve Ophthalmic Lenses | 209 |
| 5. Contact Lenses | 214 |
| *Exercises* | 216 |
| *References* | 217 |
| XIII. PRINCIPLES OF CERTAIN OPHTHALMIC DEVICES | 218 |
| 1. Ophthalmoscopy | 218 |
| 2. Retinoscopy (Skiametry) | 220 |
| 3. Badal Principle | 226 |
| 4. The Lensometer (Vertometer and Other Devices) | 227 |
| 5. The Stereoscope | 229 |
| 6. The Oculometer (Stigmatoscope) | 232 |
| 7. The Scheiner Test Method | 234 |
| 8. Biomicroscopy of the Ocular Fundus | 236 |
| *References* | 236 |
| XIV. ILLUMINATION | 239 |
| 1. Source Intensity | 240 |
| 2. Illuminance | 242 |
| 3. Luminance | 243 |
| 4. Retinal Illuminance | 244 |
| A. f-Number | 247 |
| B. Lighting Standards | 248 |

| Chapter | Page |
|---|---|
| *Exercises* | 249 |
| *References* | 250 |
| *Appendix A.* Recommended Reading | 251 |
| *Appendix B.* Answers to the Exercises | 253 |
| *Index* | 259 |

# OPTICS
## An Introduction for Ophthalmologists

## Chapter I

# PHYSICAL CHARACTERISTICS OF LIGHT

### 1. LIGHT

The current definition of light is based upon a psychophysical concept. Light is not to be identified with radiant energy alone, nor is it to be identified with the visual experience alone. "Light is that aspect of radiant energy of which the human observer is aware through the visual sensations which arise from the stimulation of the retina of the eye by those radiations."[2,7] This definition implies a definite relationship associating physical qualities of a stimulus with the visual sensations. Thus, a source of radiant energy in the physical world (radiator) corresponds to a source of luminous energy (luminator) in the psychophysical situation.

There is understandably a tendency, however, in speaking and in writing, to stray from this strict definition; we find ourselves discussing ultraviolet and infrared "light," although the radiant energy in these parts of the spectrum cannot be seen by the human eye. In this chapter we shall be concerned primarily with the physical aspects of light.

### 2. NATURE OF LIGHT

There is ample evidence that ancient man knew many of the facts about light. He learned to associate light and the transmission of light energy with heat because of the obvious relationship in the case of sunlight and the light from fire. Seeing objects in space was clearly the result, not of something that originated in the eye itself, but rather of the fact that the eye is the receptor of radiations transmitted to the eye from illuminated objects in space. These primeval men knew about the propagation of light in straight lines and about the facts of reflection. They must also have known something about the refraction of light at the surface between air and water.

All the facts known were explained on the theory that light consists of very tiny particles (corpuscles) emitted by light sources, which were believed to travel in straight lines and to pass through or bounce off objects. Some of these particles reached the eye and were absorbed, and a visual experience resulted. The speed of these particles was at first believed to be infinite, a belief held until the seventeenth century, even by the great scientists of that era.

A wave theory proposed by Huygens in 1678 was not considered seriously at the time, largely because of the influence of Newton, who adhered to the corpuscular view. Furthermore, many of the optical phenomena later explained only by the wave theory had not yet been discovered. These phenomena—diffraction, interference, polarization, and the measurements of the velocity of light in substances other than air—showed the corpuscular theory to be untenable. Similarly, in recent years, phenomena have still been found which cannot be explained on the basis of the wave theory, either.

## 3. VELOCITY OF LIGHT

The velocity of light has by now been measured by many different methods.

The first was employed about 1676 when a Danish astronomer, Römer, studied the period of revolution of a satellite of the planet Jupiter, using the newly invented telescope. At one time of the year the period, which was about 42.5 hours, seemed to be increasing, whereas at another time it seemed to be decreasing. These changes occurred in a cyclic manner as the earth completed one trip in its own orbit. The extreme difference in period was found to be twenty-two minutes. Römer concluded that this difference was due to the finite velocity of light—and that the differences represented the time required for the light to traverse the diameter of the orbit of the earth. The distance of the sun from the earth was not well known at that time, and, moreover, the accuracy of chronometers used then was far less than that of the ones used today. However, the value for the velocity of light calculated on the basis of this theory was not too greatly different from that now known.

More recent methods have given increasingly accurate values.[5] The speed of light should be remembered—namely, 186,000 miles per second, or $3 \times 10^{10}$ cm per sec. A more precise value is $2.9977 \times 10^{10}$ cm per sec. The velocity of light is now recognized as one of the most universal of the constants of nature, and this constant appears in much of the theory of atomic physics.

### 4. DISPERSION

When a narrow beam of white light is passed through a large prism, it can be observed that, in addition to the direction of the emergent light being deviated, the white light is also broken into light of different colors, the totality of which is called the *visual spectrum* (Fig. 1). This phenomenon, the breaking up of light into its component colors, is known as *dispersion*. The hues in the order in which they are increasingly deviated by the prism are red, orange, yellow, green, blue, and violet. Newton concluded from this prism experiment that white light is made up of light of colored components, because the addition of an identical prism oriented in a reversed position will recombine the colors into a white transmitted light. The change in the quality of the light in the spectrum of white light is continuous from blue to red. The

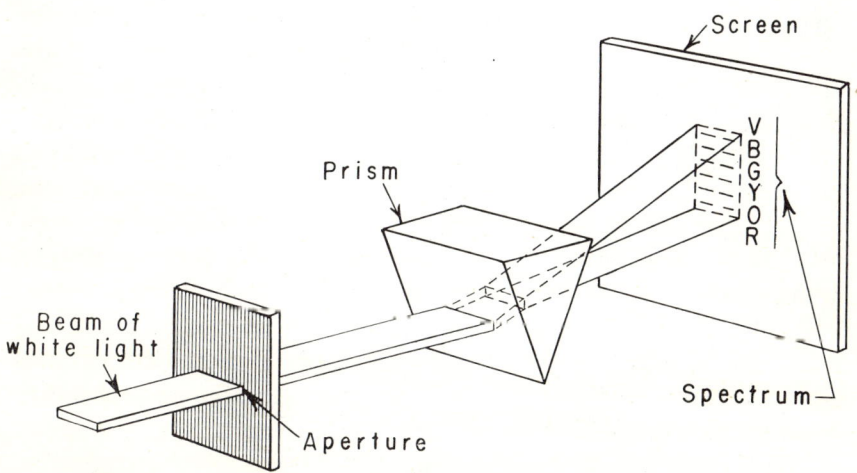

FIGURE 1. Dispersion of light by a prism.

eye identifies those portions of the spectrum with names which have been associated with them, according to color experiences. Any very small portion of the spectrum when isolated is said to be *monochromatic*.

## 5. PHYSICAL AND GEOMETRIC OPTICS

Physical optics, as the words imply, is the broad study of the physics of light. This means the study is concerned with the basic nature of light, its properties and its behavior, especially in its interaction with matter. It was here that the theory that light was a form of wave motion had its success in explaining precisely certain useful optical phenomena.

Geometric optics pertains to an area of optics by which optical problems can be solved on the basis that light can be represented by "rays" that travel in straight lines. These rays obey certain strict geometric rules in passing through optical media and when encountering interfaces between optical media. The rays of geometric optics, according to the wave theory, are but the lines of direction of propagation of wave fronts. Most, but not all, of our problems in ophthalmic optics can be solved by the rules of geometric optics.

From one point of view, physical optics pertains to the more microscopic aspects of optics, whereas geometric optics is satisfactory where apertures and distances are large compared to the wavelength of the light. These two aspects of optics are frequently made clear from the diagrams shown in Figure 2. In these, the luminous energy arises from a point source, passes through an aperture, and falls on a screen. When large apertures are used, the area of light on the screen is definitely determined by the angle subtended at the light source by the size of the aperture. Even if the aperture is reduced in size, the geometry still determines the area of light on the screen. If, however, the aperture is made quite small, then outside the bright central spot there will be found also a light pattern falling over a considerable area of the screen. This type of experiment shows that the light, on passing through the small hole, actually spreads out behind the aperture—that is, "bends around corners," a phenomenon called *diffraction*. Thus,

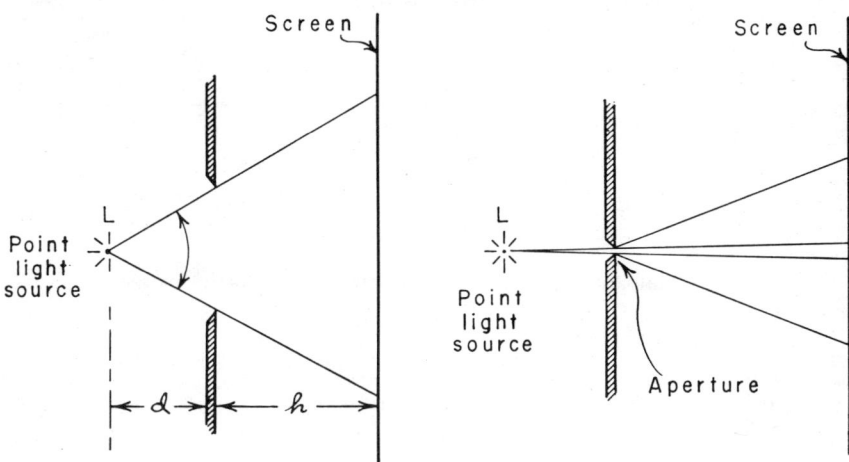

FIGURE 2. Scheme illustrating the difference between geometric and physical optics in terms of size of aperture.

the simple geometric theory whereby light is represented as traveling strictly in straight lines fails to describe the behavior of light passing through small apertures. The resulting light pattern is explainable by physical optics and, in particular, by a wave theory of light. Actually, evidence of diffraction will be found even when larger apertures are employed, if the edge of the light pattern on the screen is carefully examined.

## 6. WAVE THEORY OF LIGHT

A wave theory of light makes possible the explanation of a number of the basic phenomena of optics that cannot easily be accounted for by any other theory. To understand light as a wave motion, we can only make use of analogies such as the behavior of waves on the surface of water.

When a series of waves move across a body of water, it can be observed that a wood chip on the surface of that water is not itself moving in the direction the wave travels. On the contrary, it and the "particles" of water at any point appear to oscillate vertically— at right angles to the direction of the wave propagation. If we were to represent graphically the successive vertical positions

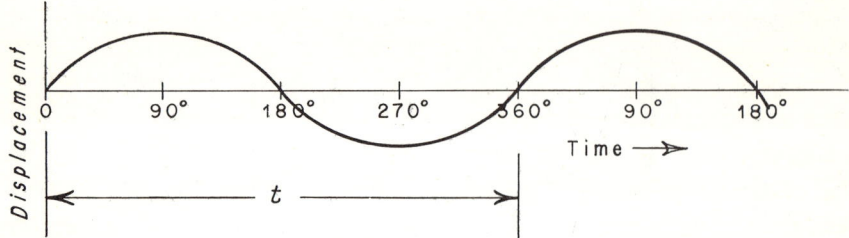

FIGURE 3. Sinusoidal trace illustrating displacement of a particle subject to wave motion.

(displacement) of a given particle at successive moments as this particle moves during the passage of the wave, we would obtain a (sinusoidal) trace as shown in Figure 3. This type of motion is said to be simple harmonic. In the time represented by $t$ the particle would have completed a "cycle" in its movement. One complete wave would have passed a given position on the surface during this time. The distance the wave has traveled on the surface of the water in the time $t$ depends upon the velocity ($V$) of the wave, and this distance equals the product of $V \cdot t$. The distance is called the *wavelength* and is usually represented by the Greek letter lambda ($\lambda$). Thus, $\lambda = V \cdot t$. If $t$ is the time, in seconds, for the completion of one cycle of the wave motion, then the number of cycles completed in one second, which is the frequency, $n$, is given by the reciprocal of $t$, $n = 1/t$. Thus, the velocity of a wave will be $V = n\lambda$.

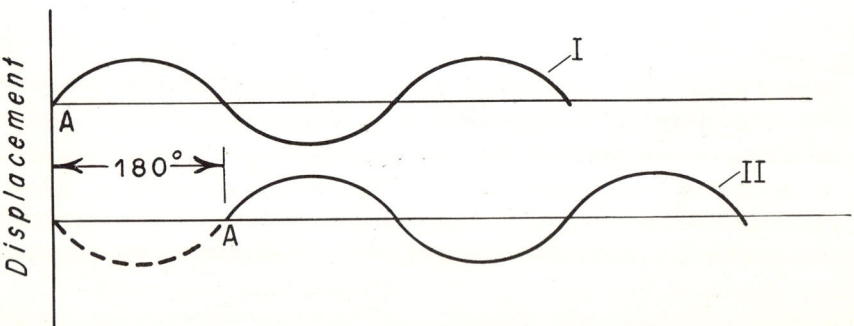

FIGURE 4. Illustration of two wave motions 180° out of phase.

Two waves of the same wavelength and frequency, but beginning their cycles at different times, are said to be *out of phase*. The idea of a cycle to describe the behavior of a wave is analogous to the motion of a particle in a circle. A cycle would be a rotation of 360°; a half-cycle, of 180°; and so on. Thus, it is customary to designate the phase difference between two waves in terms of a proportion of a cycle, expressed in arc degrees. If two identical waves impinging upon the same particle are out of phase 180° (Fig. 4), at any moment the displacement of the particle due to one wave would be equal to, but in a direction opposite to, the displacement due to the second wave. The resultant displacement would therefore be zero, and the wave motion would be destroyed. It is by such analogy that one explains the experimental fact of light interference. On the other hand, if these two wave motions are *in phase* (phase difference is zero), one obtains an enhancement of the wave amplitude.

The wave theory of light assumes that there are vibrations of some kind giving rise to the wave motion that are at right angles to the direction of the propagation. Huygens assumed that every point on a wave front acts as a new source, sending out waves of the same frequency and amplitude. The envelope of all the new "wavelets" would constitute the new wave front, and all points on this front would have the same phase. The original theory of Huygens had to be modified in some respects before it could adequately explain optical phenomena.

## 7. INTERFERENCE

The phenomenon of interference can be demonstrated by an experiment (attributed to Young) schematically illustrated in Figure 5. Suppose the light from a distant monochromatic source falls on a very narrow slit, $S$, in an opaque screen. The light spreading out from this slit, because of diffraction, then passes through two other very narrow, closely spaced slits in a second screen. From each of these slits the light also spreads out because of diffraction, and the light from the two slits overlaps. If a lens is placed beyond these slits to focus the light on a third screen, one will find on that screen a series of parallel alternating light-

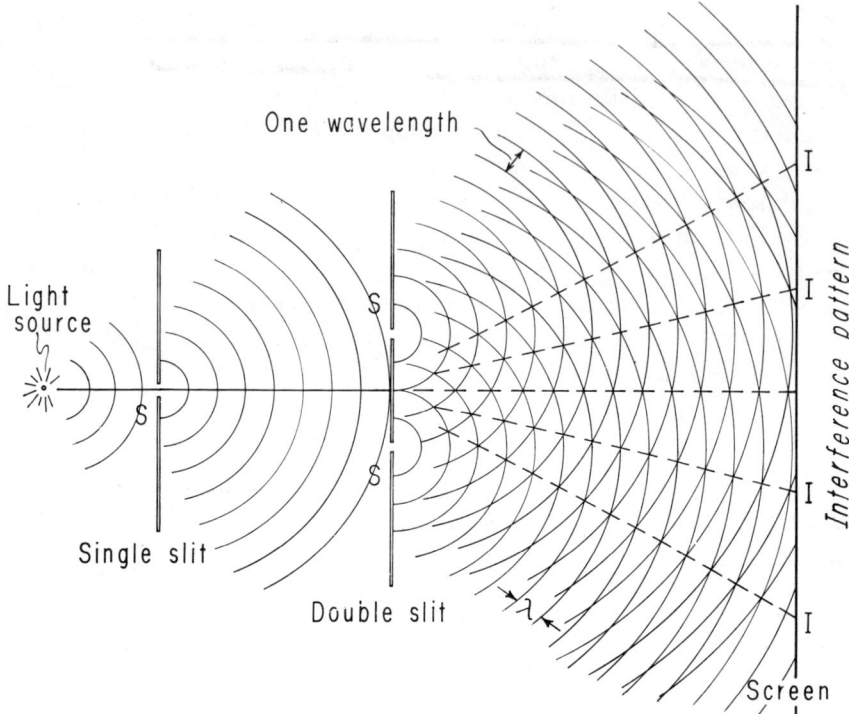

FIGURE 5. Essential scheme of experiment for illustrating interference of light.

and-dark lines, the brightness of the lines slowly decreasing away from the center of the array. Here, because of the phenomenon of interference, the two sets of waves from the two slits augment or destroy each other, depending upon the phase difference that results from the difference in distances of the two slits from points on the third screen. The intervals of no light indicate those points at which the wave series from the two slits are destroyed by interference; the light intervals denote those points at which these waves augment each other by addition of their amplitudes. When the difference in distance of a point on the screen from the two slits results in phase differences of 180°, destructive interference occurs.

The colored bands seen in thin films of oil on the surface of water (called *Newton's fringes*) also are results of the phenomenon of interference due to a variation of thickness in the oil film. Figure 6 illustrates the physics involved. When the eye accom-

## Physical Characteristics of Light 11

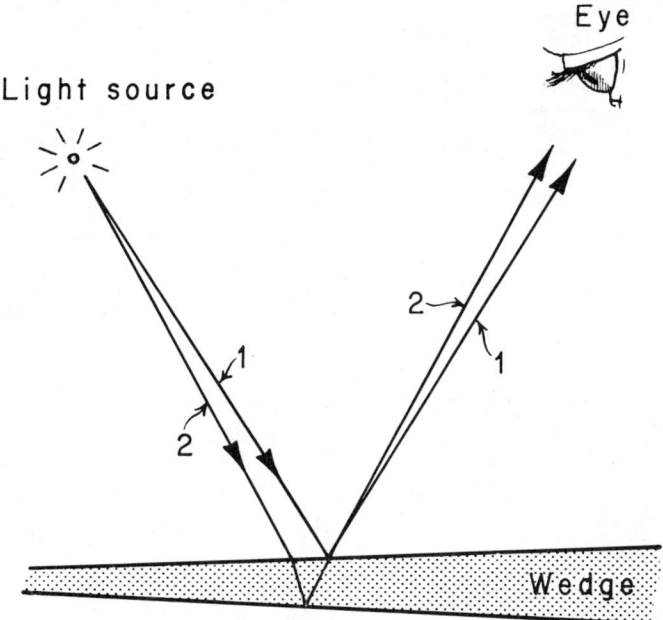

FIGURE 6. Conditions favorable for the formation of Newton's fringes.

modates for the surface of an optical wedge, rays reflected from the first and from the second surfaces are focused on the same point on the retina. The light from these two interfaces differs in phase corresponding to the difference in lengths of the optical paths. This difference increases with the increase in depth between the interfaces. Thus a pattern is formed of light fringes where the two wave fronts are in phase (phase difference is zero) and dark fringes where they differ in phase by 180 degrees. The change in depth between the interfaces can be calculated if the wavelength of the incident light is known and the separation between successive light (or dark) fringes is measured.

If one observes in monochromatic light the reflections from the interfaces of two optical surfaces consisting of a plane and a sphere, or of two spheres with different radii of curvature, one can observe a series of concentric rings about the point of contact. The separation of the rings is related to the difference in curvature of the two surfaces. The accuracy in the curvature of optical surfaces can

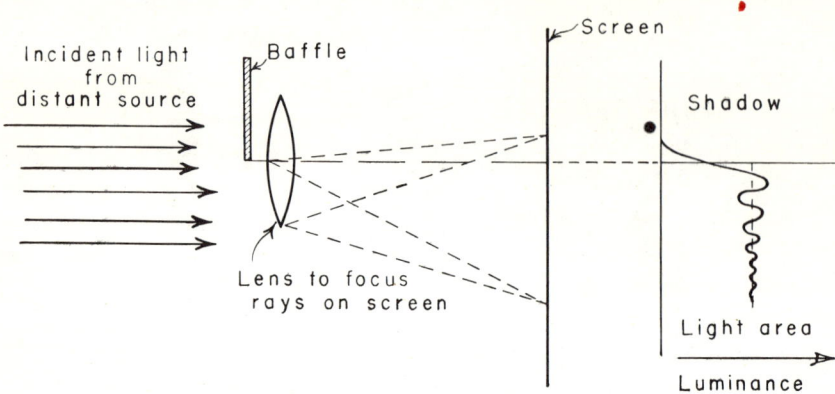

Figure 7. Diffraction of light from the edge of a baffle.

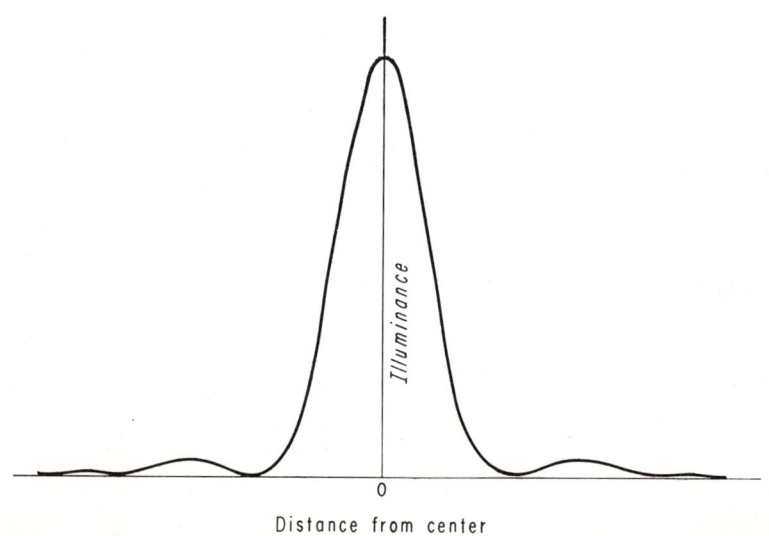

Figure 8. Distribution of illuminance in an image, due solely to diffraction.[3] (Modified by permission of Interscience Publishers.)

be tested by using Newton's rings, since these rings or fringes appear as the contours of a topographical map. This method of surface testing is used routinely in precision optical shops. When white, instead of monochromatic, light is employed, these rings are colored because color is associated with different wavelengths.

## 8. DIFFRACTION PATTERNS

Although the term *diffraction* applies to the fact that light actually bends around or spreads out from corners, as a phenomenon diffraction occurs only when the light passes by the edge of an opaque object. Diffraction phenomena occur, therefore, throughout optics and especially in optical instruments in which light is limited by apertures. When light passes an opaque edge and is allowed to fall on a screen, a diffraction pattern can be seen on the screen near the edge of the shadow. This pattern is the result of interference of wavelets originating near the edge of the obstruction or baffle. If the baffle has a straight edge (Fig. 7), the diffraction pattern in the vicinity of the geometric shadow will consist of a gradation of light intensity into the shadow area and a series of parallel light and less light lines of decreasing difference in brightness and of decreasing separation into the illuminated area.

Most lens systems have circular apertures, and the diffraction pattern caused by this circular boundary is important in the theory of image formation. Figure 8 illustrates the distribution of the intensity of light in the image due solely to diffraction. Such a distribution forms a circular pattern of a relatively bright central disk, surrounded by a succession of dark and light rings, the intensities of the light rings decreasing rapidly. This pattern is known as Airy's disk. The central area of this pattern often sets the limit of how small the image can be.

## 9. DIFFRACTION PATTERNS AND THE DIFFRACTION GRATING

Suppose we have two very narrow, slightly separated parallel slits in an opaque screen and that we illuminate these by monochromatic light from a distant source (Fig. 9). The light passing through these slits is focused upon a second screen by a suitable lens. On that screen can be seen a characteristic diffraction pattern consisting of a bright central image on each side of which there are a number of separated fainter lines. These fainter lines appear at the places on the screen where the diffracted waves from the two slits are in phase according to the concept of interference. The angular position, $\theta$, of these lines from the central image is related to the separation of the slits, $s$, and the wavelength of the light, $\lambda$. For example, at the moment a given wavelet arrives at the center of the slit $S_2$, there will have been positions of waves having the same phase at distances $1\lambda$, $2\lambda$, and so forth, from the second slit. Since waves having the same phase constitute a wave front, there will be a series of wave fronts proceeding at different directions from that of the central image. Shown in the figure is the wave front originating from $S_2$ and the point on an arc (not drawn) one wavelength ($\lambda$) distant from $S_1$, which is proceeding in the direction indicated by $\theta$. From the little right triangle, of which the slit separation $s$ is the hypotenuse and the adjacent side represents the particular wave front, we can write that $s \sin \theta = \lambda$. This relationship determines the position of the bright line at $P$ on the screen, and this line is said to be a *first-order* spectral line. If we used the wave front arising from $S_2$ and the point on an arc $2\lambda$ from the slit $S_1$, there would be another bright line on the screen still farther from the central image, a line called the *second-order* spectral line. Thus, in general $s \sin \theta_n = n\lambda$, in which $n$ is the spectral order. These lines of the diffraction pattern are at the same distances on each side of the central image.

The use of only two slits gives diffraction lines, but their intensities are much too weak to be useful. To increase the intensity of the diffraction lines, the number of slits is greatly increased, to form what is called a *diffraction grating*. Gratings can be made on glass for a transmission grating or on a highly polished reflecting material for a reflection grating. The slits consist of grooves or lines engraved by a sharp stylus by a specially designed ruling engine, which will rule many thousands of identical parallel grooves per centimeter. In order to obviate the use of a lens with the transmission grating, as shown in the figure, the lines are often ruled upon a slightly concave reflecting surface, resulting in what is called a *concave* diffraction grating. This, then, not only serves as a grating but at the same time focuses the light on a screen for visual observation or on suitably situated photographic plates.

The position of $P$ on the screen (Fig. 9) will be different for different wavelengths, so for nonmonochromatic incident light there will be a spectrum or a series of bright lines of different hue on the screen for each of the orders (Fig. 10). The order of the

## Physical Characteristics of Light

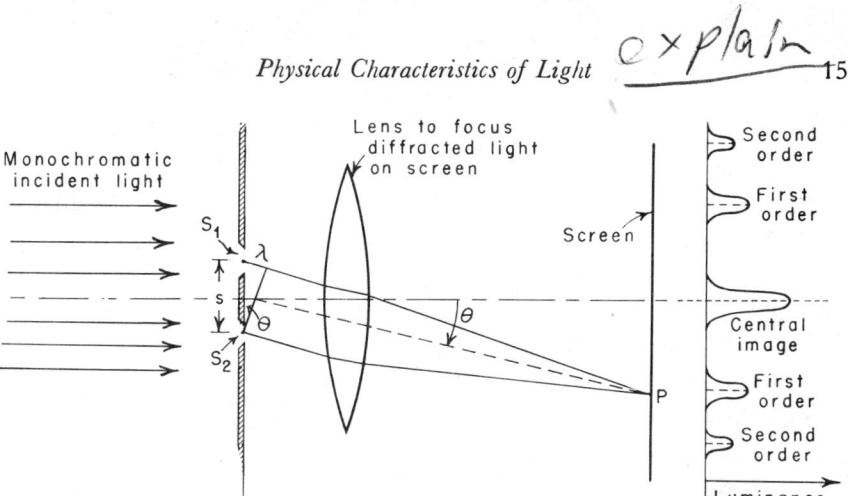

FIGURE 9. Scheme illustrating the theory of the diffraction grating.

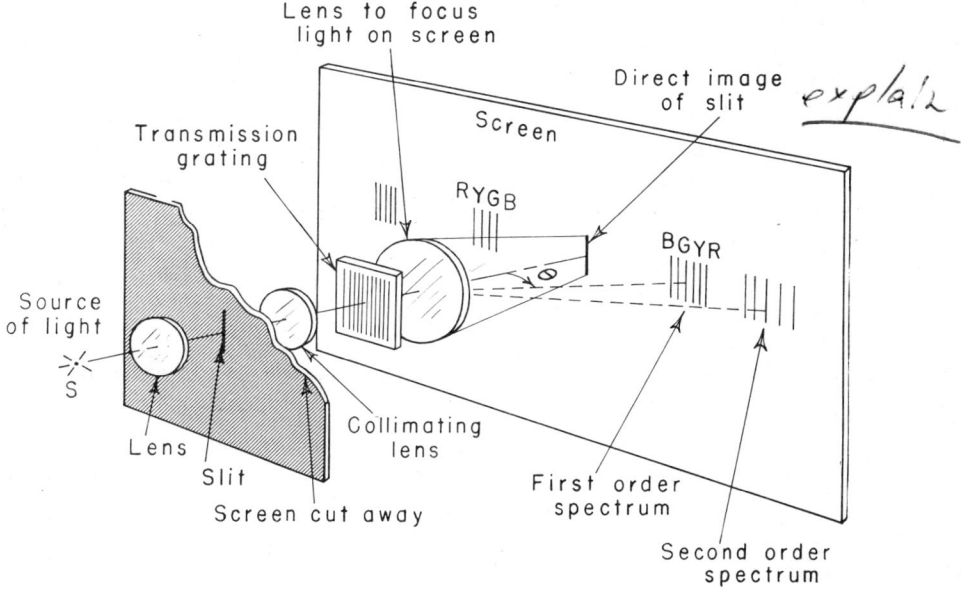

FIGURE 10. Essentials of experimental setup for producing diffraction spectra.

hues in the spectrum formed by a grating will be reversed from that produced by a prism; the *red* will be deviated more than the *blue*. For higher orders the spectrum of one may overlap that of the other. The diffraction grating is essentially the basic tool for determining the wavelength of light, for one needs only to know accurately the separation of the slits and to measure the angle $\theta$.

Wavelengths are expressed in various units based on the metric system: microns = $\mu$ = $1/1000$ = $10^{-3}$ mm; millimicrons = $m\mu$ = $10^{-6}$ mm, or Angstrom units (Å) = $10^{-7}$ mm. We shall use millimicrons ($m\mu$). The visible portion of the spectrum varies from about 400 $m\mu$ (violet) to about 700 $m\mu$ (deep red).

Two types of spectra are distinguished:

1. Continuous spectra that arise from metals heated to incandescence.
2. Line spectra that arise from an arc or a spark between substances or through gases. Every element has a separate and distinct line spectrum, and each element can be identified by its particular spectral pattern.

For optical purposes it has been convenient to refer to three separate portions of the spectrum by the wavelengths of three of the so-called Fraunhofer lines. These are dark lines seen against a continuous spectral background of the light from the sun and are known to be caused by the absorption of light at these wavelengths by certain cooler gases surrounding the sun. The ones most often used in ophthalmic optics are as follows:

```
        C...............656 mµ (red) (hydrogen)
        D...............589 mµ (yellow) (sodium)
        F...............486 mµ (blue) (hydrogen)
```

As will be seen later, the indices of refraction of glass are usually stated for spectral light at these three wavelengths.

For the purposes of this introduction to optics, the discussion here is merely to indicate briefly how the wave theory of light explains these phenomena and at the same time to show how the wavelengths of light can actually be determined.

## 10. POLARIZATION OF LIGHT

Under certain circumstances it can be shown that there is a lack of uniformity in the character of a light beam in different meridians perpendicular to the direction of propagation. Under these circumstances, the light is said to be *polarized*. This phenomenon can be explained only if light waves occur by virtue of a transverse vibration—that is, perpendicular to the direction of propagation. The light reflected from the surface of any substance (except metallic surfaces) is at least partially polarized. The light rays transmitted by certain substances found in nature, such as birefringent crystals, also are polarized.

The wave theory states that in normal propagation of light the transverse vibrations are equal in all meridians. The vibrations of light reflected from the surfaces of nonmetallic substances tend to be all in one meridian—that is, plane-polarized; those vibrations concerned with the light transmitted tend to be in the plane at right angles to that reflected. The plane of polarization of reflected light is said (arbitrarily) to be perpendicular to the surface. Other more useful ways of producing polarized light are with the Nicol prism and with sheet Polaroid.® The Nicol prism consists of a birefringent calcite crystal so cut that the transmitted ray is entirely plane-polarized (the other ray polarized at right angles is totally reflected and absorbed in the prism mounting). Polaroid consists of a suspension of iodide crystals in a transparent plastic sheet. In the manufacture, this sheet is uniformly stretched in one direction, which then orients all the crystals in that direction. About 80 per cent of the light transmitted is polarized in one plane, and less than 1 per cent is transmitted in the plane at right angles. Whether a given beam of light is polarized can be tested only by rotating a Nicol prism or a Polaroid plate (the analyzer) before the beam and observing the change in the intensity of the transmitted light. If this intensity passes through maxima and minima as the analyzer is rotated, then the light is at least partially polarized. If, at certain positions, the light is entirely extinguished the incident light is plane-polarized.

Polaroid plates used before the eyes are useful in reducing reflected glare from surfaces, such as that from water. Of course, the intensity of the useful light transmitted is reduced at least by 50 per cent. Certain transparent substances, including glass, under stress affect the polarization of the light passing through them. Thus, the presence of instability or strain in an ophthalmic lens can be detected if, when the lens is placed between crossed Nicol prisms or Polaroid sheets, an irregular light pattern is observed. By the use of Polaroid film and vectographs, studies on

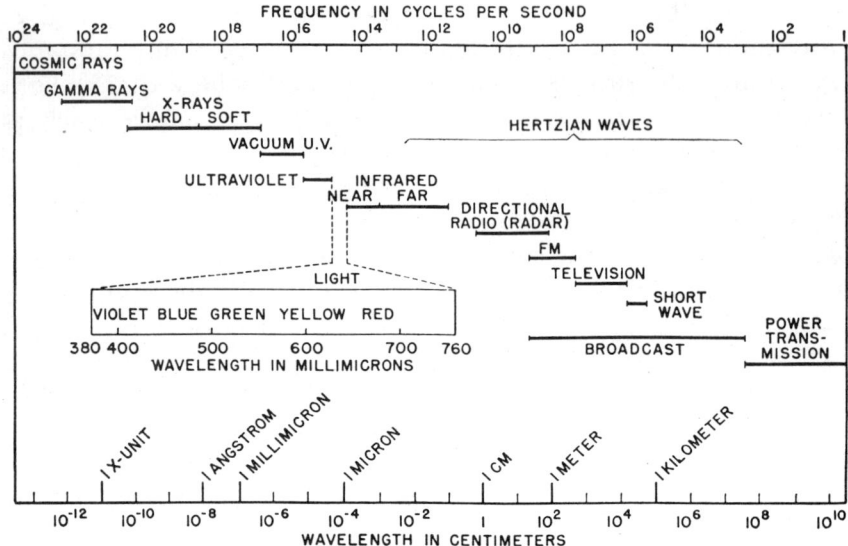

FIGURE 11. The electromagnetic spectrum.[6] (By permission of the Illuminating Engineering Society.)

binocular vision are simplified because different images can be exposed separately to each of the two eyes from the same target.

Two Polaroid plates mounted parallel, one behind the other, will function as a continuous light attenuator if one of them can be rotated on their common axis. When the planes of polarization are in the same meridian, a maximal amount of light is transmitted, $I_0$; when at right angles, no light is transmitted. If the angle between the planes of polarization is $\theta$, the intensity of the light transmitted would be given by $I = I_0 \cos^2 \theta$ (law of Malus), in which $I_0$ is the maximal transmission—that is, when the planes of polarization are in the same direction and $\theta$ is zero.

## 11. ELECTROMAGNETIC THEORY OF LIGHT

In 1864 Maxwell developed a theory that the vibrating particles postulated in the wave theory actually are moving electric charges and that the propagated wave motion itself is a magnetic radiation. This theory is mathematical in exposition and is difficult to visualize. In 1887 Hertz experimentally verified the existence of electromagnetic radiations from oscillating electric disturbances. This experiment was the origin of wireless telegraphy and provided the evidence that light radiations are, in fact, electromagnetic. Today, light is considered to be but one small portion of a large spectrum of electromagnetic radiations, from the extremely short gamma rays through the x-rays—ultraviolet—visible—infrared—Hertzian waves—radar—radio, and so on. Figure 11 illustrates the electromagnetic spectrum. The range of wavelengths to which the eye responds is indeed only a small portion of the entire spectrum.

## 12. QUANTUM OPTICS

Just as the phenomena of diffraction, interference, and polarization cannot be accounted for by a corpuscular theory, there are other phenomena in optics that cannot be explained by a wave theory. One of these is the photoelectric effect, in which it is found that electrons are ejected from a substance when it is irradiated. Similarly unexplainable by a wave theory are the phenomena of the spectral energy of radiation sources, the existence of line spectra, the behavior of spectral lines when the radiation source is placed in a magnetic field, and so forth. These phenomena have been explained with amazing precision by the quantum theory. This theory assumes that radiant energy consists of packets or quanta of energy, the magnitudes of which are proportional to the frequency of the radiation. This is to say, the energy $E$ in each quantum is given by $E = hc/\lambda$, in which $c$ is the velocity of light, $\lambda$ is the wavelength associated with the quantum (both $c$ and $\lambda$ being in the same metric units), and $h$ is a constant known as Planck's constant (equal to $6.54 \times 10^{-27}$ erg/sec). The energy in a quantum of radiant energy of a long wavelength would be less than that in a quantum of radiant energy of a short wavelength.

## 13. SPECTRAL ENERGY OF RADIATION SOURCES

The concept of "black-body" radiation is important in optics. A perfect black body is one which will absorb equally all the energy of radiations that fall upon it, regardless of wavelength. Similarly, such a body when heated will emit radiations whose

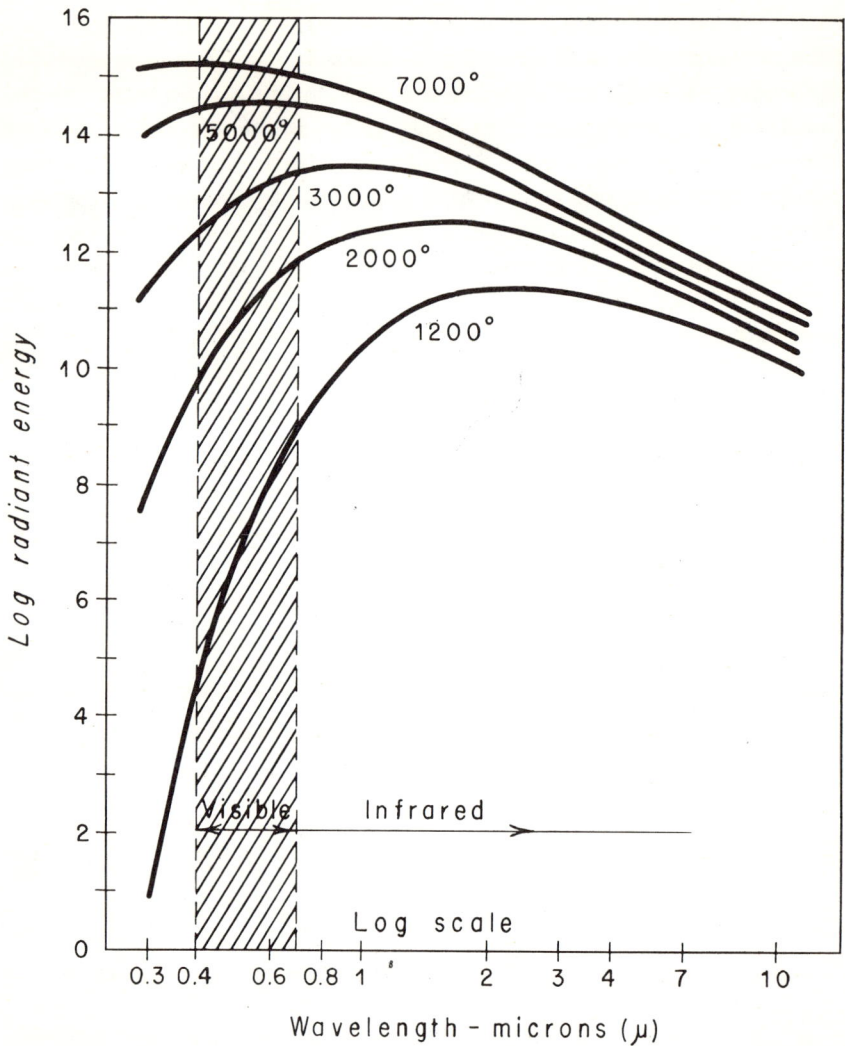

FIGURE 12. Spectral distribution of radiant energy from a black body heated to various temperatures.

spectral distribution depends, not on any characteristic of the body itself, but only on the temperature.

If the intensity of the radiation from a heated black body is measured at every wavelength of the spectrum by special instruments, a characteristic distribution of energy intensity for each temperature will be found. These distributions are shown schematically by the graph in Figure 12.[4] At low temperatures the radiation is entirely in the infrared and is not visible. As the temperature is increased, the radiations begin to be visible, but the greatest part of the emitted energy is still predominately in the red. As the temperature is further increased, a greater and greater proportion of the energy of the emitted radiation occurs in the blue and violet portions of the visible spectrum and thus becomes more intense and more "bluish."

Most surfaces and common light sources are fairly nonselective in the visible portion of the spectrum; therefore they can be approximately matched in chromaticity by light from a complete radiator at some temperature. It has become usual, then, to describe the spectral composition of the light emitted or reflected from a surface in terms of a *color temperature*. Examples are as follows:

| | |
|---|---|
| Old-style carbon filament lamp | 2080° K (absolute) |
| 100-watt tungsten filament lamp | 2830° K ,, |
| Photoflood lamp | 3430° K ,, |
| Carbon arc | 3700° K ,, |
| Daylight plus sky light | 6500° K ,, |
| Standard daylight source C (chromaticity diagram) | 6740° K ,, |
| North sky, clear day | 25,000° K ,, |

Thus, color temperature pertains not to an actual temperature but to that temperature of a black-body radiator which would yield the same distribution of intensity in the visible portion of the spectrum.

## 14. INTERACTION OF LIGHT WITH AN OBJECT

Whenever light falls upon an object four things may occur. First, the object may transmit part of the light energy falling on it. In such a case we say that the object is *transparent* or is *translucent*. The light transmitted by certain substances (calcite) is all or partially plane-polarized. If none of the incident light is transmitted, the object is said to be *opaque*.

Second, all or a portion of the incident light may be reflected. The light reflected from an object may also be selective as to wavelength, and this accounts for the color of the object. If the

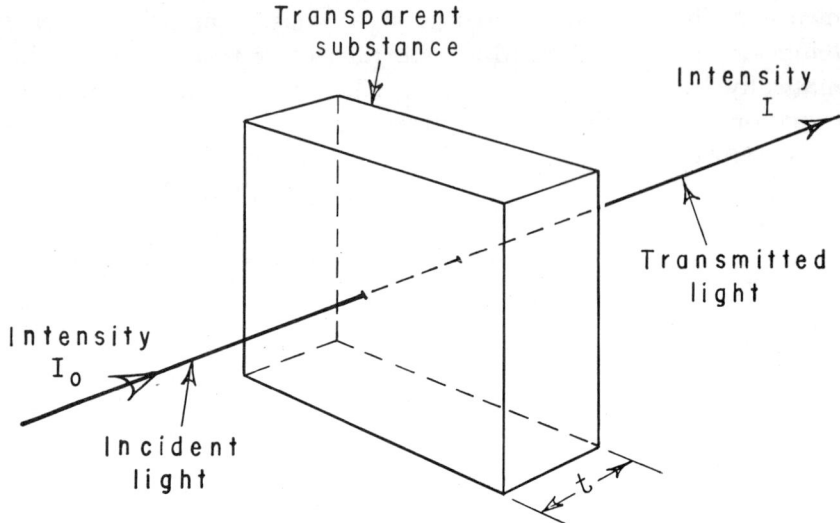

FIGURE 13. Illustration of the transmission of light through substances.

object is very small compared to the wavelength of light, the light *reflected* is said to be *scattered*, and the angle of scattering measured from the angle of incidence will vary according to the wavelength of the light, the blue being scattered through the larger angle. This accounts for the blue sky and the red-yellow sunsets—scattering of the light by molecules of water vapor and dust in the atmosphere.

Third, all or a portion of the incident light may be absorbed. When light is absorbed, the energy (1) is transformed into heat, which increases the temperature of the body or is dissipated; (2) is converted into chemical energy, with some change in structure; (3) is reemitted in the form of light of longer wavelength; and (4) ejects electrons.

Finally, the light falling on a surface exerts a pressure on that surface.

## 15. ABSORPTION

It frequently is necessary to refer to the absorption properties of various materials (sunglasses, filters and the like); hence some general idea of the terminology customarily used is important.

When a beam of light is incident to a plane-parallel piece of a transparent homogeneous substance, part of the light will be transmitted and the rest absorbed (and reflected). The useful beam is that transmitted. If $I_o$ indicates the intensity of the light incident to the plate (Fig. 13) and $I$ is the intensity of the transmitted light, then the *transmission*, $T$, of the plate would be the ratio $T = I/I_o$. This ratio is always less than one and is frequently expressed as percentage of transmission. The *opacity*, $O$, of the plate would be the reciprocal of the transmission: $O = 1/T = I_o/I$.

If several such plates of different transmissions were used in succession, the final transmission would be the product of the separate transmissions $(T_1 \times T_2 \times T_3 \ldots)$, or the resultant opacity would be the product of the separate opacities $(O_1 \times O_2 \times O_3 \ldots)$. If, now, one writes the transmission of a given plate as $T = 10^{-D}$, or the opacity $O = 10^D$, in which $D$ is defined as the *density*, then the final transmission of a series of plates would be $10^{-(D_1+D_2+\ldots)}$, or the opacity would be $10^{+(D_1+D_2+\ldots)}$. Thus $D$ is a useful notation. It is defined specifically as $D = \log_{10} O = \log_{10}(1/T) = \log_{10}(I_o/I)$ (the common logarithm or that to the base 10). The total density of a series of plates would be the sum of the separate densities.

Examples of transmissions, equivalent opacities, and densities are given below.

|  | % | 50% | 10% | 5% | 1% | 0.25% | 0.1% | 0.01% |
|---|---|---|---|---|---|---|---|---|
| Transmission | $T$ | 1/2 | 1/10 | 1/20 | 1/100 | 1/400 | 1/1000 | 1/10,000 |
| Opacity | $O$ | 2/1 | 10/1 | 20/1 | 100/1 | 400/1 | 1000/1 | 10,000/1 |
| Density | $D$ | 0.3 | 1.00 | 1.30 | 2.00 | 2.60 | 3.00 | 4.00 |

Sunglasses, filters, and so on may be specified by the percentage of transmission or by the density.

The transmission of light through a substance may be spectrally selective. By this is meant that the transmission varies according to the wavelength. All substances are more or less selective in some parts of the spectrum. The so-called *neutral-tint* filters are specifically designed to have nearly the same transmission for all wavelengths of the visible spectrum. The Wratten gelatin filters, Corning glass filters, and interference filters, on the other hand, are specifically devised to be highly selective in different parts of the spectrum so that they may be used to transmit light of some dominant hue.

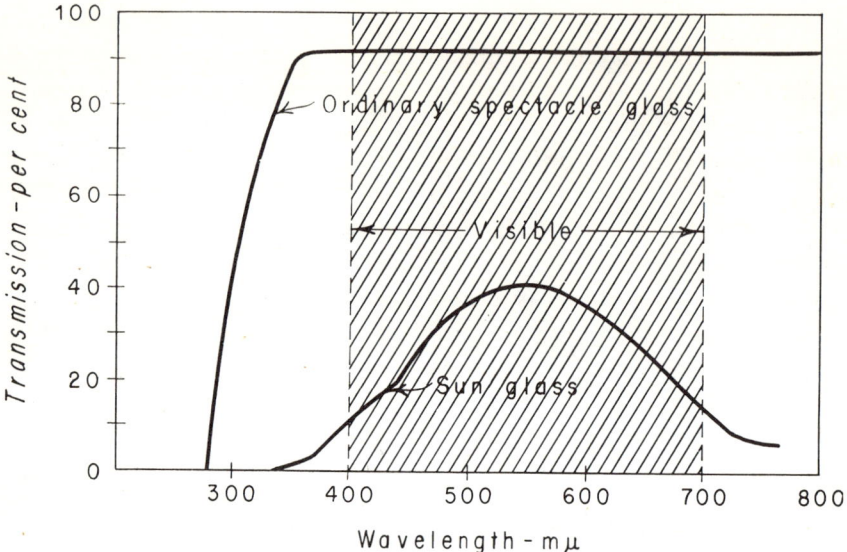

FIGURE 14. Approximate spectral transmission of light through spectacle glass and one type of sunglass.

The spectral transmission curves of ordinary spectacle glass and of glass used for sunglasses are approximately those shown in Figure 14. The transmission curve for spectacle glass shows that ultraviolet light shorter than 300 m$\mu$ is almost completely absorbed.

The approximate spectral transmission of the refractive media of the young-adult human eye is illustrated in Figure 15.[1] The curve shows that ultraviolet radiation below 300 m$\mu$ does not reach the retina, while a considerable range of infrared rays do reach the retina.

Mention also should be made here of Bouguer's (or Lambert's) law because the discussion about transmission, absorption, and density really follows from that law. When light of a certain intensity $I_o$ is incident to a substance, the intensity $I_x$, because of absorption, at a distance $x$ from the plane of incidence, is given by $I_x = I_o e^{-ax}$, in which $e$ is the base for natural logarithms (2.71828) and $a$ is the *coefficient of absorption* of the substance. This law states that the intensity of the transmitted light decreases exponentially with the thickness and with the coefficient of absorption of the

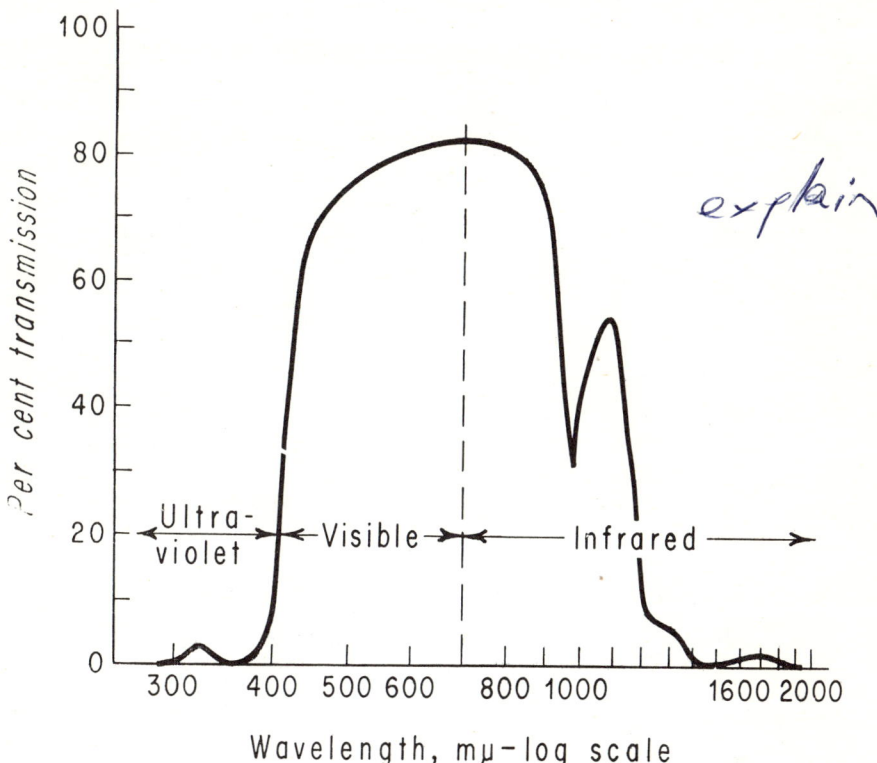

FIGURE 15. Spectral transmission of light by the ocular media of a young-adult human eye.[1] (Modified by permission of the C. V. Mosby Company.)

substance. This law is sometimes used to describe the transmission of solutions, in which case it is known as Beer's law, *a* being replaced by *bc*, in which *c* is the concentration of the solute and *b* is called the *molecular absorption coefficient* (or the *extinction coefficient*).

## REFERENCES

1. BOETTNER, E. A., and WOLTER, J. R.: Transmission of the ocular media. *Invest Ophthal*, 1:776-783, 1962.
2. COMMITTEE ON COLORIMETRY, OPTICAL SOCIETY OF AMERICA: *The Science of Color*. New York, Crowell, 1953, p. 220.
3. DITCHBURN, R. W.: *Light*, 2nd ed. New York, Interscience, 1963, p. 201.
4. FOWLE, F. E.: Radiation from a perfect (black body) radiator. In Washburn,

E. W.: *International Critical Tables of Numerical Data—Physics, Chemistry, and Technology.* New York, McGraw, 1929, vol. 5, pp. 238-242.
5. GRAY, D. E.: *American Institute of Physics Handbook.* New York, McGraw, 1957, p. 119.
6. ILLUMINATING ENGINEERING SOCIETY: *IES Lighting Handbook*, 3rd ed. New York, Illuminating Engineering Society, 1962, p. 1-2.
7. JONES, L. A.: The historical background and evolution of the colorimetry report. *J Opt Soc Amer, 33:*534-543, 1943.

*Chapter II*

# GEOMETRIC OPTICS—REFLECTION

## 1. PENCILS, RAYS, BEAMS OF LIGHT

For descriptive purposes in geometric optics, a selected portion of the light from a luminous source is sometimes called a *pencil* of light (Fig. 16). Such a pencil of light would be delimited by one small hole, $A_1$, and one large hole, $A_2$. A pencil of light may also be thought of as consisting of a bundle of *rays*, each ray being a straight line aimed in a particular direction from the light source and all of them—including adjacent rays in a pencil— diverging from the source. A ray of light as defined could pass through only two exceedingly small holes. A *beam*, on the other hand, would pass through two large apertures, and the rays in the beam would be more or less parallel. Obviously, a ray is a hypothetical concept but nevertheless a very useful one, for a ray will indicate the direction of propagation of the light energy in a pencil or beam. At any point the ray is perpendicular to the wave front.

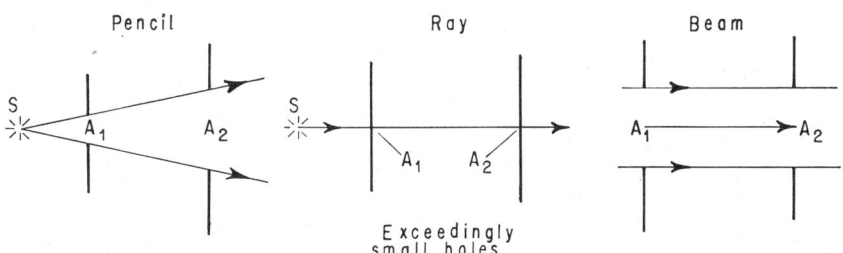

Figure 16. Illustrations of the definitions of a pencil, ray, and beam of light.

## 2. THE CONCEPT OF OPTICAL IMAGES

It is important to have a clear understanding of the concept of optical images and the formation of optical images. Suppose we

28    OPTICS: An Introduction for Ophthalmologists

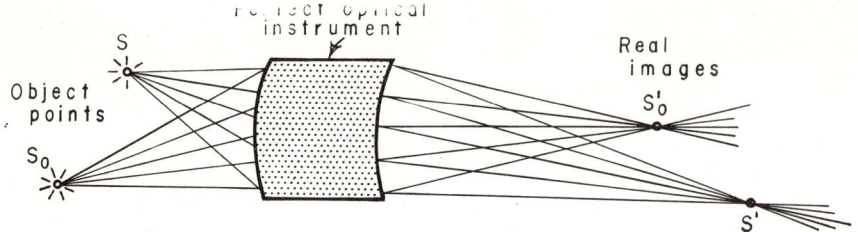

FIGURE 17. Principle of image formation by an ideal optical system producing real images.

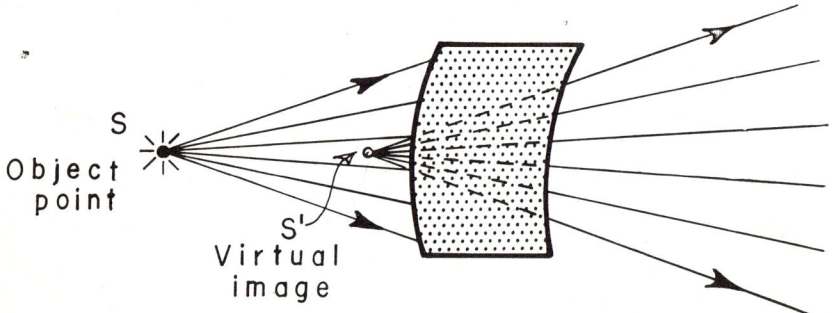

FIGURE 18. Principle of image formation by an ideal optical system producing virtual images.

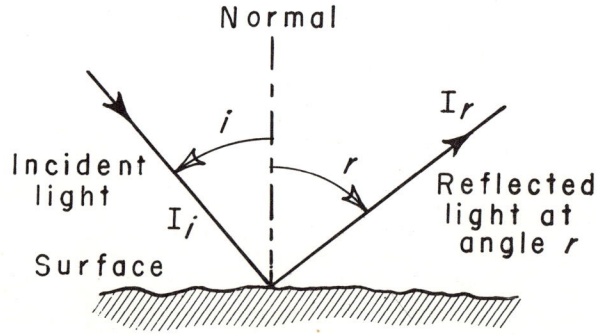

FIGURE 19. Reflection from a mat surface.

have a *perfect* optical system (Fig. 17), perhaps consisting of lenses, prisms, mirrors, and so on. When *all* the rays of light from a point $S_o$ (the object point) on passing through the system converge to a point $S_o'$ and continue beyond, $S_o'$ is said to be the *real* image of $S_o$. It is real because the rays from the object actually converge to an image. A photographic film at that position would be exposed by the light of the image. If the rays emerging from the optical system diverge, rather than converge (Fig. 18), *as if* originating from a point on the object side of the system, that point $S'$ is called a *virtual* image. The image formation is said to be *stigmatic* (from the Greek, meaning "point") when all the rays from the point source on passing through the system converge to a point. If, for every position of a light source, $S$, *all* the rays converge to a corresponding image point $S'$, the optical system is said to be *aplanatic*.

The positions of points $S_o$ and $S_o'$ (or $S$ and $S'$) are called *conjugate points*. In an optical system any pair of conjugate points are *reversible*, in that the positions of object and image can be interchanged.

The extent to which different rays from $S$, on passing through the optical system, do not intersect at the same point $S'$ is said to be due to *aberrations*. The different kinds of aberrations will be discussed more fully later.

## 3. REFLECTION OF LIGHT

The characteristics of the light reflected from a surface depend upon the texture of the surface. If that surface is smooth or highly polished, the reflection is *specular* or *regular*, as in the case of a mirror. If the surface is rough or mat, the reflection is *diffuse* because the light from an incident beam is reflected in many different directions. In this case the luminous intensity of the reflected light may be greater in some directions than in others, depending upon the surface texture and upon the direction of the incident light. The surface reflecting the light may also be spectrally selective; that is, the luminous intensity of the reflected light may vary with the wavelength of the incident light.

If light of a certain luminous intensity $I_i$ is incident on an unpolished surface (Fig. 19) and if $I_r$ is the luminous intensity of the reflected light, the *reflectance*, $R$, of the surface is defined as $R = I_r/I_i$. This ratio is always less than one because some of the

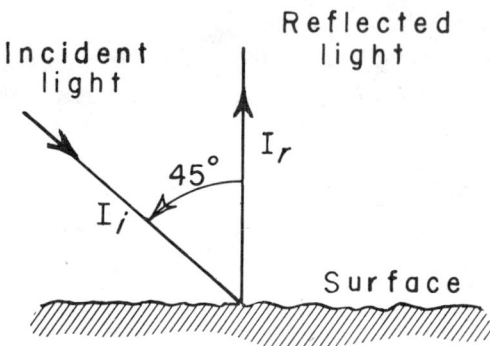

FIGURE 20. Standard relationships for the directions of the incident and the reflected light in the measurement of reflectance of a surface.

incident light energy is always absorbed, and the intensity of the reflected light is always less than that of the incident light. The reflectance, $R$, sometimes, if not usually, varies with the angle of incidence and with the angle at which the intensity of the reflected light is measured. For standardization of the reflectance of mat surfaces, it is recommended that the angle of the incident light be 45° and the luminous intensity of the reflected light be measured normal (perpendicular) to the surface (Fig. 20). The reflectance of a very white mat surface may be of the order of 0.80, or 80 per cent.

## The Law of Reflection

Consider first the reflection from a highly polished surface—namely, that of a *plane* mirror. The principles of the wave theory may be used in deriving the fundamental law of the specular reflection from such a surface, in that wave fronts will be considered perpendicular to the direction of propagation of the light. The locus of points at which the vibrations are in phase constitutes the wave front.

Suppose a narrow beam of light from a very distant source falls upon the surface at an angle $\theta$ (theta) measured between the direction of propagation and the surface (Fig. 21). Consider now

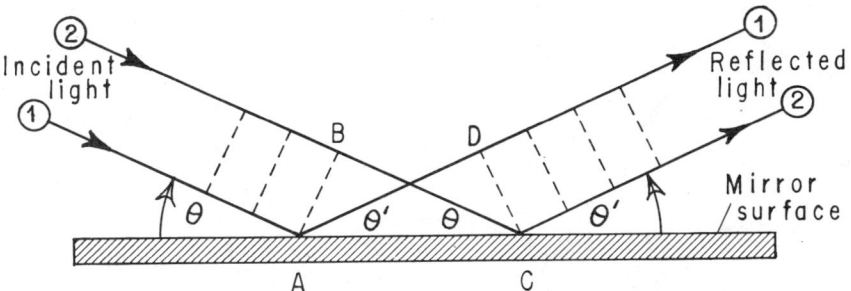

FIGURE 21. Diagram used in deriving the law of reflection of light from a plane mirror.

any two rays of the incident beam—say, rays 1 and 2. These rays are parallel because the luminous light source is so far away. There are a series of plane wave fronts propagated in the direction of incidence. At the moment ray 1 arrives at the surface at point $A$, the wave front can be represented by $AB$. The point $A$ will act as a new light source, and ray 1 will be reflected in a certain direction. Now, the point $B$ on ray 2 will continue in the same direction of incidence until it also arrives at the surface at the point $C$. During the time that the wave front at $B$ has traveled to $C$, the wave front at $A$ will have proceeded a distance $AD = BC$. Then $CD$ is the new wave front after reflection, just as $AB$ was the wave front before reflection. The wave front $CD$ will now travel in the direction at right angles to itself.

Let the angle of reflection be $\theta'$. From inspection of the figure, then, the two right triangles $ABC$ and $CDA$ can be identified. Since the hypotenuse of each of these triangles is common to both and the distance $AD = BC$, the two triangles are congruent and the included angles are equal: $\theta = \theta'$. This is the law of reflection: *The angle of incidence is equal to the angle of reflection.* The incident and the reflected rays lie in the same plane.

It is standard practice, in geometric optics, always to specify the angle of incidence, $i$, and the angle of reflection, $r$, as measured from the normal, that is, from the line perpendicular to the surface (Fig. 22). The law of reflection of individual rays is useful in describing the formation of images by reflection.

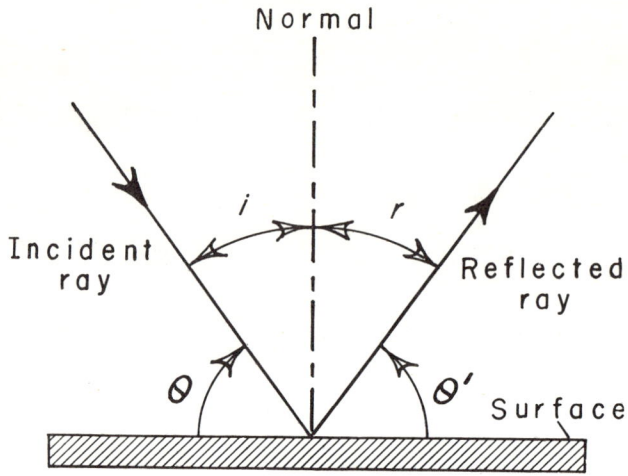

FIGURE 22. Standard method of designating angles of incidence and of reflection.

## 4. IMAGE FORMATION BY REFLECTION
### A. The Plane Mirror

By use of the law of reflection, the relationship between the distances of the object and image can be derived in the reflection from the surfaces of mirrors. Consider first the plane mirror. Suppose there is a light source $S$ a distance $h$ from the surface of a plane mirror (Fig. 23). Consider *any* ray, say, $SB$, from $S$ that strikes the surface at $B$. This ray makes an angle of incidence $i$, measured from the normal at $B$. This ray $SB$ is reflected at the same angle $(r)$ and proceeds in the direction $BC$. Consider now the ray $SP$ that travels perpendicularly to the surface of the mirror. The angle of incidence is zero, and therefore the angle of reflection is zero; that is to say, the ray travels back upon itself. The rays $BC$ and $SP$ are reflected, then, as if they came from the point of intersection of their extensions, the point $S'$. Inspection of this figure shows that the right triangles $SPB$ and $S'PB$ must be congruent because the line segment $PB$ is common to both and the opposite angles are equal $(i = r)$. Therefore, $SP = PS' = h$. Since the ray $SB$ is any ray, all rays from $S$ will be reflected as if originating from the point $S'$, and $S'$ is therefore the virtual image of $S$. The image $S'$ will always lie behind the mirror on the line perpendicular to the mirror through $S$, and its distance from the

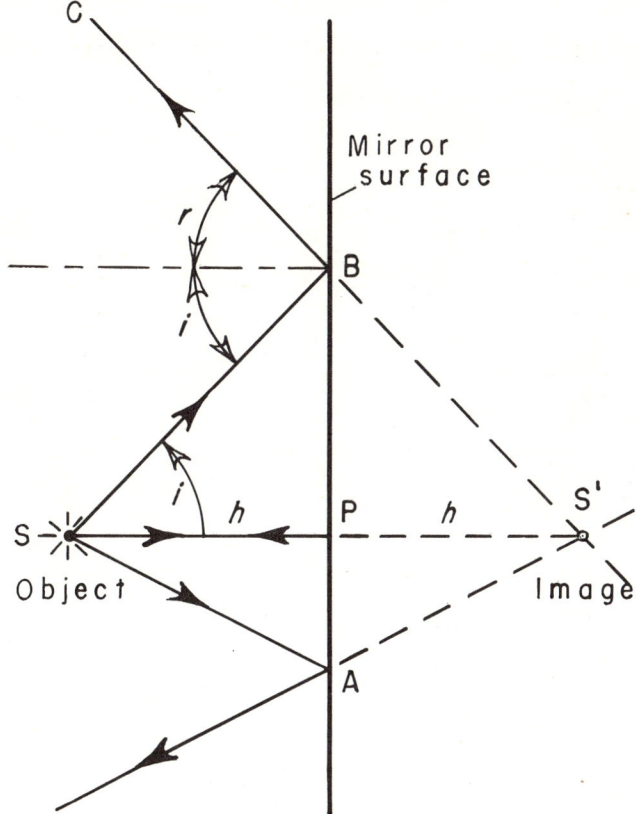

FIGURE 23. Image formation by a plane mirror.

surface will be equal to the distance of the object $S$ in front of the mirror.

The particular rays utilized by the eye in observing the image reflected from a plane mirror can be readily determined by use of the construction just outlined. Suppose the object before the mirror is an arrow $AB$ (Fig. 24), and the image $A'B'$ is observed by an eye at the position illustrated. The position of the image can be immediately located by drawing perpendiculars from $A$ and $B$ to the mirror surface and by placing the image points $A'$ and $B'$ behind the mirror surface the same distances that $A$ and $B$ are in front of the mirror. Now, there is only one central ray from each of the images of $A$ and $B$ to the eye; these are the rays that strike the

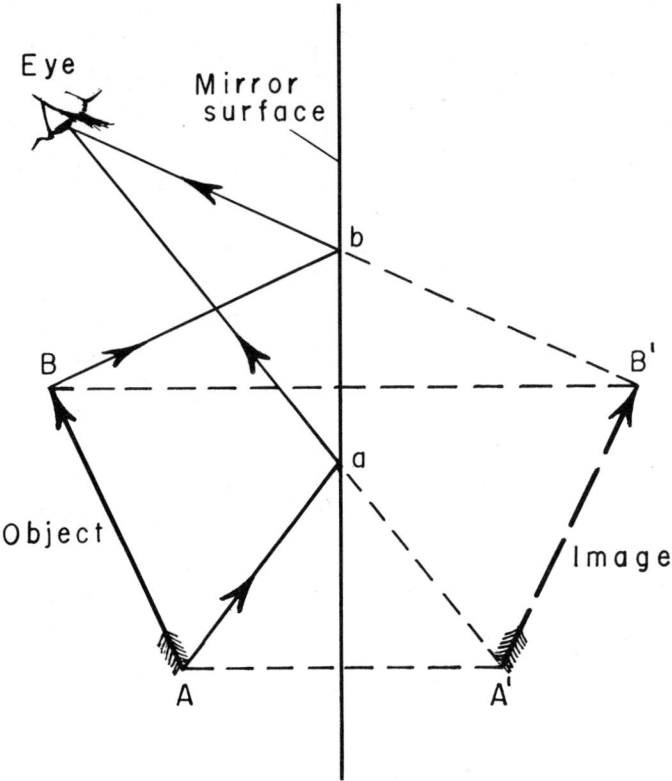

FIGURE 24. Method of diagramming rays to the eye from a reflected image.

surface of the mirror at *a* and *b*. Thus, only the portion of the mirror between *a* and *b* is used by the eye in observing the entire image.

## B. Spherical Mirrors

There are many examples in ophthalmology in which convex and concave reflecting surfaces play important roles. Consequently, the optical imagery obtained from such surfaces is an important topic. The rays from an object point before such a surface are reflected so that an image of the point is formed. The general manner in which individual rays are reflected to form the image is illustrated for theoretically ideal convex and concave mirrors in Figures 25 and 26, respectively. In the first case (Fig. 25), the image is *virtual,* since the reflected rays diverge as if originating

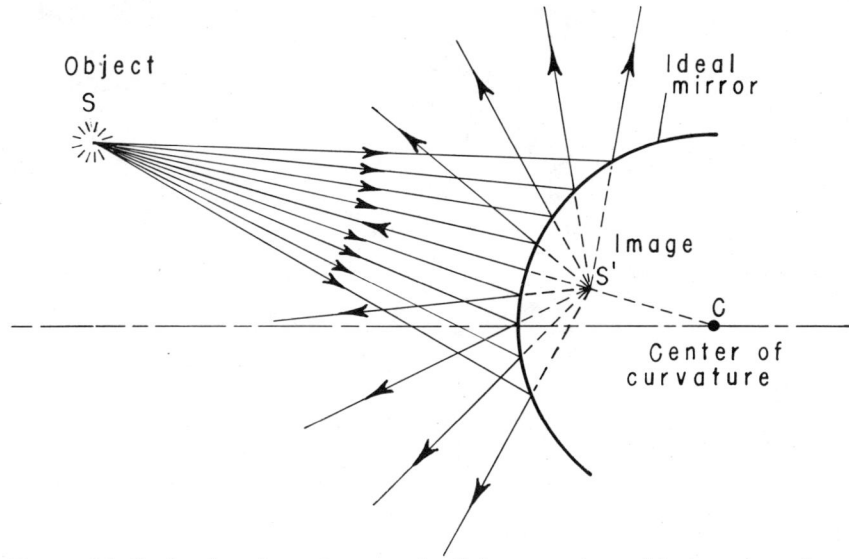

FIGURE 25. Reflection of rays from a point-light source by an ideal convex mirror.

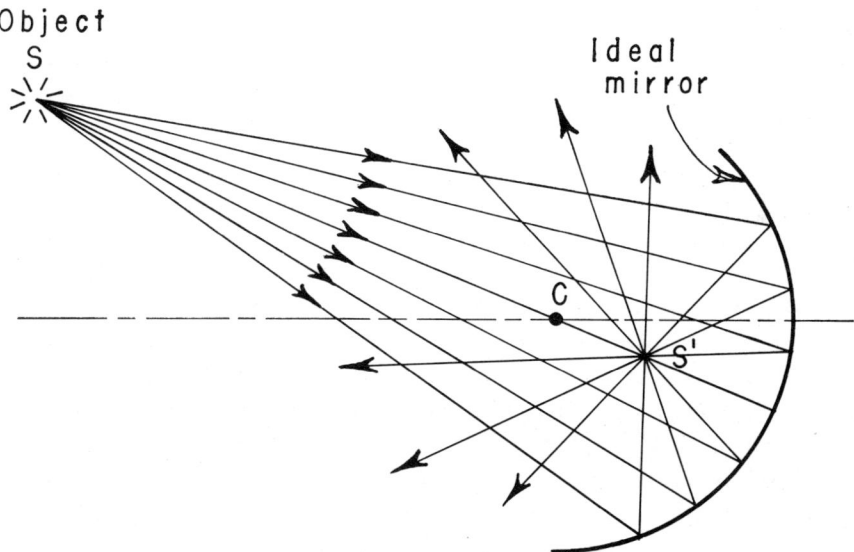

FIGURE 26. Reflection of rays from a point-light source by an ideal concave mirror.

from a point $S'$ behind the surface. In the second case (Fig. 26), the image $S'$ of the object point is *real*, for the reflected rays actually converge to a point.

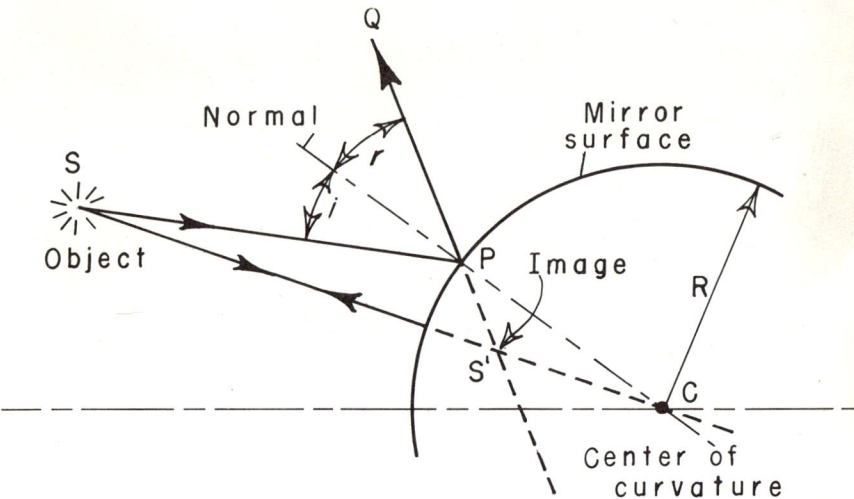

FIGURE 27. Diagram illustrating the method of locating the image of a point by an ideal convex mirror.

### 1. The Convex Mirror.

Consider now, in more detail, the convex mirror with a spherical surface. In Figure 27, suppose $S$ is an object point in front of the surface whose center of curvature is at $C$ and whose radius of curvature is $R$. Here again is employed the construction procedure that was used to locate the image in the plane mirror—namely, to select several rays from $S$ that fall on the surface and to find where the reflected rays intersect. Consider then, any ray—say, $SP$. By drawing the line from the center of curvature $C$ to $P$, the normal to the surface is established, and then the angle of incidence, $i$, can be measured and, correspondingly, the angle of reflection, $r$, can be laid out. The particular ray from $S$ to the center of rotation of the surface, $SC$, is useful, for this ray coincides with the normal or perpendicular to the surface; hence it is reflected back on itself. The intersection of the prolonged rays $SC$ and $PQ$ would be a point at $S'$, and this point would be the position of the image of the two rays selected. This image is a virtual image, for the two reflected rays $CS$ and $S'Q$ are diverging as if from the point $S'$.

To discuss the problem in more quantitative terms, suppose there is an object defined by the points $A$ and $B$ in front of the convex mirror (Fig. 28). We are interested in determining the

# Geometric Optics—Reflection

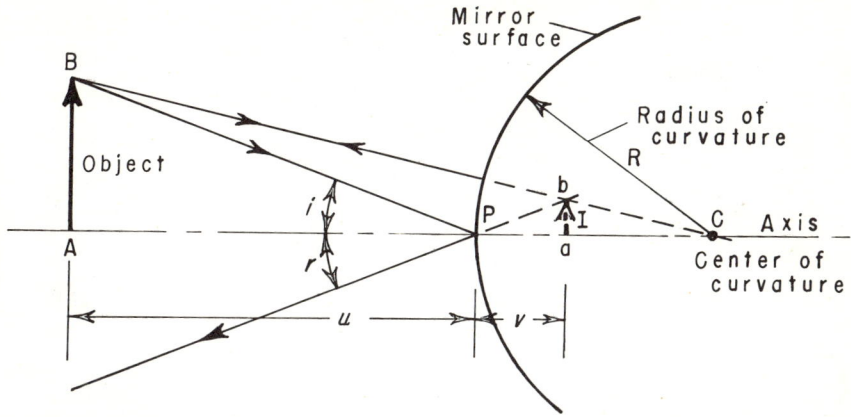

FIGURE 28. Diagram used in deriving the relationship between the distances of the object and the image in the reflection by a convex mirror.

positions of their images $a$ and $b$, and the relative size $ab$ in terms of the radius of curvature and location of $AB$ from the mirror. The radius of curvature is $R$. Let us measure all distances from the pole, $P$, of the surface, whence $u$ is the distance of the object $AB$ from the surface, and $v$ is the distance of the image $ab$ from the surface. Let the ray from $A$ directed to the center of curvature be the "optical axis," and let $AB$ be perpendicular to that axis. Of all the rays that leave the object $B$, let us select two: one aimed to the center of curvature and the other aimed to the pole $P$ of the surface. The first ray is reflected back on itself, and the second is reflected at the same angle as the angle of incidence. Inspection of the figure made up of these rays shows that there are two similar triangles included, the sides of which are proportional. These triangles are shown isolated in Figure 29. From the triangles $ABC$ and $abC$, we can write

$$\frac{ab}{AB} = \frac{R - v}{R + u}. \qquad \ldots (1)$$

From the triangles $ABP$ and $Pba$, which are right triangles with the same included angle, since $r = i$, we can write also

$$\frac{ab}{AB} = \frac{v}{u}. \qquad \ldots (2)$$

This equation relates the sizes of the image and object to their distances from the reflecting surface. Equating (1) and (2),

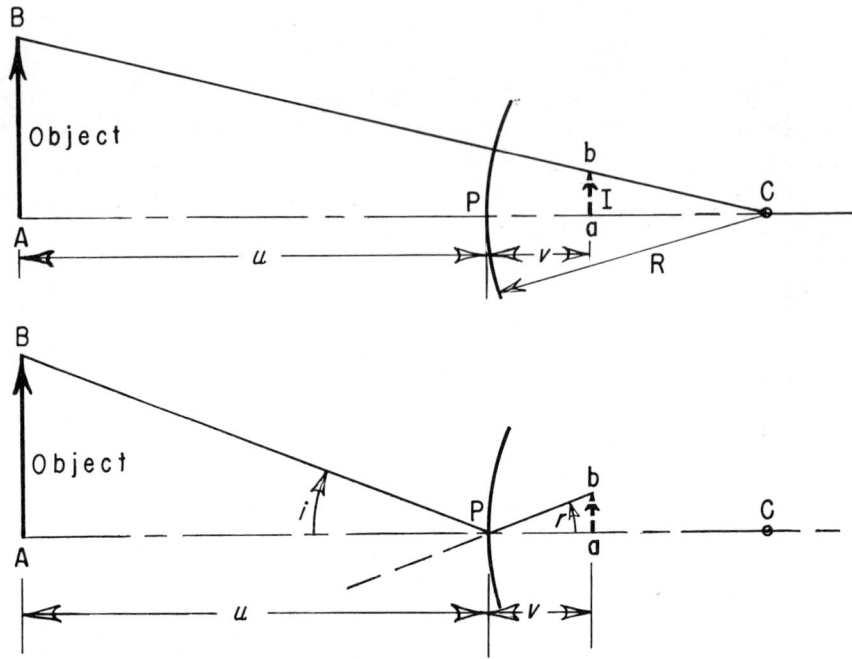

FIGURE 29. Particular similar triangles isolated from Figure 28 for special emphasis in derivation.

$$\frac{R-v}{R+u} = \frac{v}{u},$$

whence

$$u - v = \frac{2}{R} u v,$$

and, finally,

$$\frac{1}{v} - \frac{1}{u} = \frac{2}{R}. \quad \ldots (3)$$

This formula, then, relates the relative distances of the image and object in terms of the radius of curvature of the convex surface. Note that if $u$ is large (that is, if the object is a great distance from the mirror), $1/u \rightarrow$ zero; hence the distance of the virtual image from the mirror is $v = \frac{1}{2}R$, or one half the radius. This distance is called the *focal length*, $f$, of the mirror.

Already equation 2 shows that the ratio of size of the image, $ab = I$, to the size of the object, $AB = O$, is $v/u$. This ratio $I/O$ is called the *magnification*.

These relationships describe quantitatively the experimental

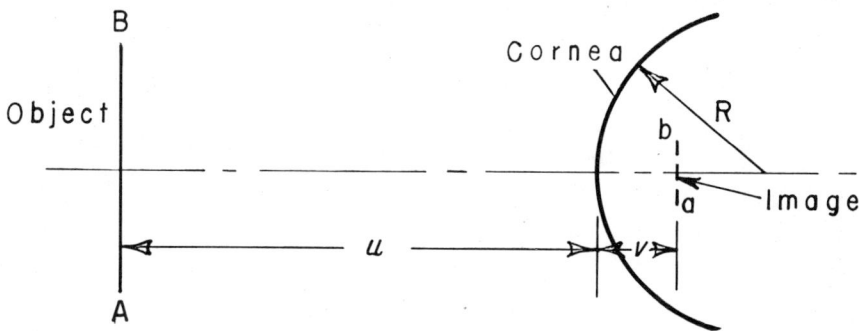

FIGURE 30. The basic principle of the ophthalmometer for measurement of the radius of the cornea of the eye.

facts of the reflection from a convex surface. The qualitative facts are (1) the image is always virtual for a real object; (2) the image is always erect; and (3) the image is always smaller than the real object.

*Example 1:* The cornea of the eye is a convex mirror, and the highlight usually seen in it is a reflected image. Suppose we have a circular object that is 20 cm in diameter—perhaps Placido's disk—located at 40 cm from the eye. If the radius of the cornea is 0.8 cm, how far is the virtual image from the pole of the cornea and what is its size? Substitute the values given into equation 3, to produce $1/v - 1/40 = 2/0.8$, from which one finds $v = 0.396$ cm $= 3.96$ mm. Because the radius is so small, the image lies essentially at one half the radius of curvature of the cornea. The size of the image in millimeters, from equation 2, would be

$$ab = (200)(3.96)/400 = 7.92/4.00 = 1.98 \text{ mm.}$$

*Example 2:* An important application of these relationships is found in the reverse problem, for if by some method the size of the reflected corneal image can be measured, and the distance and the size of the object are known, the radius (and therefore the power) of the cornea can be calculated. This is the principle of the ophthalmometer (Fig. 30). We use these two equations:

$$1/v - 1/u = 2/R, \text{ and} \qquad \ldots \text{(a)}$$
$$ab/AB = I/O = v/u. \qquad \ldots \text{(b)}$$

Elimination of $v$ is effected by solving equation b for $1/v$ and substituting this value in equation a. Then the radius of the cornea, $R$, is readily found to be $R = 2uI/(O - I)$. Usually, however, the size of the object $O$ is so much larger than that of the image $I$ that $O - I = O$. Thus

$$R = 2uI/O. \qquad \ldots \text{(4)}$$

In the ophthalmometer the size of the image ($I$ or $ab$) is measured by an optical *doubling* device. The object or *mire* size ($O$ or $AB$) remains the

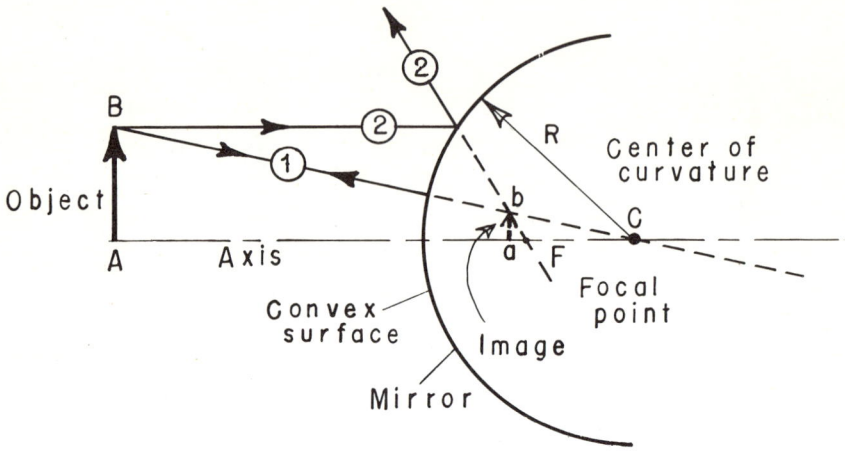

FIGURE 31. Simplified method of diagramming the position and size of the image in the reflection by a convex mirror.

same, and the distance of the object is kept constant in the focusing of the telescope. Thus the radius of curvature of the cornea is proportional to the measured size of the reflected image.

Placido's disk, consisting of a series of bright concentric rings, is useful for detecting irregularities in the curvature of the corneal surface. The corneal reflex images of these rings are observed through a small hole in the center of the disk. Distortions in the appearance of these images indicate irregularities in the curvature of the cornea, and certain distortions may indicate the presence of keratoconus.

When the distance of the object from the vertex of the mirror is very great, the image would be at a distance $½R$. This image is said to lie in the focal plane of the mirror at $F$, and this plane would be at the focal length or focal distance $f = ½R$ from the vertex of the mirror. With the concept of the focal point, the ray diagrams useful to determine qualitatively the position of an image can be simplified (Fig. 31). We need only to draw two rays from the object point $B$—one through the center of curvature which is reflected back along the same ray, and the other parallel to the axis $AC$. This latter ray on reaching the mirror is reflected as if originating at the focal point, $F$, of the mirror. The point of intersection of these two rays determines the position of the image, $b$.

Geometric Optics—Reflection 41

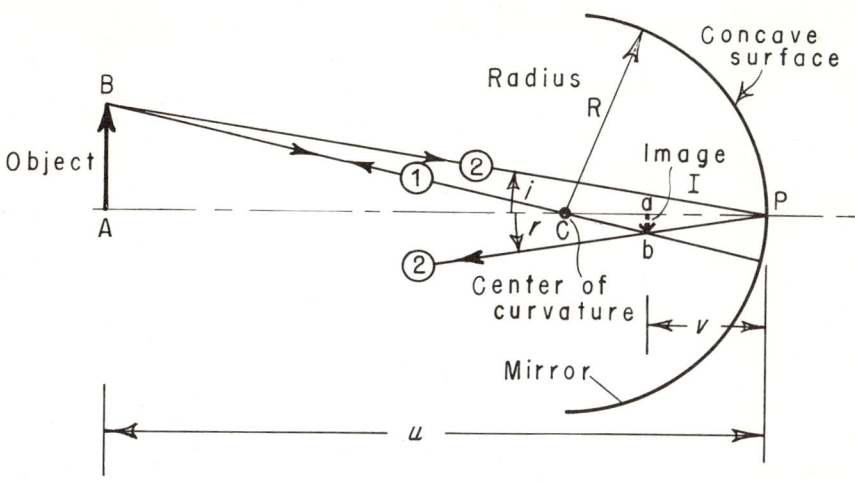

FIGURE 32. Diagram used in deriving the relationship between the distances of the object and the image in the reflection by a concave mirror.

## 2. The Concave Mirror.

For a concave surface, the quantitative relationship between the distances of the object and image, and that between their sizes, can be derived by following the same procedure as that used for convex surfaces.

In Figure 32, let $AB$ again be an object, the rays from which are reflected from the concave surface of radius $R$. The ray from $A$ directed through the center of curvature, $C$, is taken as the optical axis. Of all the rays that leave the object point $B$, two particular rays are selected that easily determine the position of the image $b$. The first ray (1) from $B$ is aimed through the center of curvature of the surface. This ray, then, is incident to the surface normally and hence will be reflected back upon itself. The second ray (2) is one that proceeds from $B$ aimed to the pole $P$ of the mirror. This ray then makes an angle of incidence $i$, since the line $AP$ is also normal to the surface of the mirror. This ray is then reflected by the mirror surface at an equal angle ($r = i$) from the same line $AP$. The point at which this reflected ray intersects ray 1 or $BC$ determines the position of the image $b$ of the object $B$. As before, let $u$ be the distance of the object $AB$ from the surface and $v$ be the distance of the image $ab$ from the surface. Now again in this diagram there are two sets of similar triangles, the sides of which are proportional.

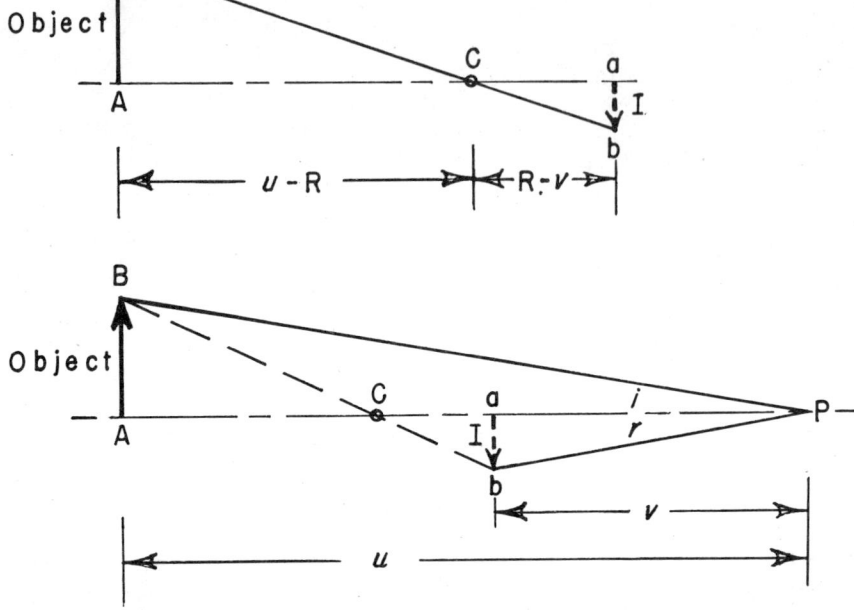

FIGURE 33. Particular similar triangles isolated from Figure 32 for special emphasis in derivation.

The first set consists of similar triangles $ABC$ and $abC$, shown isolated in Figure 33. From these

$$\frac{ab}{AB} = \frac{I}{O} = \frac{R-v}{u-R}. \qquad \ldots (5)$$

The second set consists of the similar triangles $ABP$ and $abP$, also shown isolated in Figure 33. From these one obtains the formula for the magnification—namely,

$$\frac{ab}{AB} = \frac{I}{O} = \frac{v}{u}, \qquad \ldots (6)$$

which expresses the relationship between the sizes of the image and the object and their distances from the reflecting surface.

These two expressions can be equated to eliminate $ab/AB$, whence

$$\frac{v}{u} = \frac{R-v}{u-R}.$$

This can be simplified to

$$\frac{1}{v} + \frac{1}{u} = \frac{2}{R}. \qquad \ldots (7)$$

With the positions of the object and image as shown in Figure 32, this equation differs from that (equation 3) for the convex mirror in that the sign of $u$ is plus instead of minus.

## 5. GENERAL RELATIONSHIPS FOR MIRRORS

The same formula can be used for both convex and concave mirrors if we assign algebraic signs to all distances and adopt the following convention:

1. All optical problems will be diagrammed so that the incident light travels from left to right. Object and image distances will be measured along the axis *from* the reflecting surface.
2. A real object will always be located to the left of the mirror surface. In some problems in which a virtual image must be used as the object to the right of the surface, that distance will be taken minus.
3. When the image is on the right side of the mirror surface, its distance will be taken plus. In some problems in which it may be placed or occurs to the left of the surface, the image distance must be taken minus.
4. If the center of curvature of the optical surface is to the right of the surface, the focal length and radii of curvature will be taken plus; if to the left of the surface, they will be taken minus.

For the convex mirror, the center of curvature and therefore the focal length will be plus and $u$ and $v$ will be plus, as shown in Figure 28. For the concave mirror (as diagrammed in Fig. 32), however, $R$ (and therefore $f$) and $v$ must be made minus. Thus we may adopt the general formula

$$\frac{1}{v} - \frac{1}{u} = \frac{2}{R} = \frac{1}{f}, \qquad \ldots (8)$$

which will hold for either the convex or the concave mirror. For the convex mirror, $R$ is taken plus; for the concave mirror, $R$ is taken minus.

The relationships between the positions of real objects and their images for the concave mirror are a little more complicated than

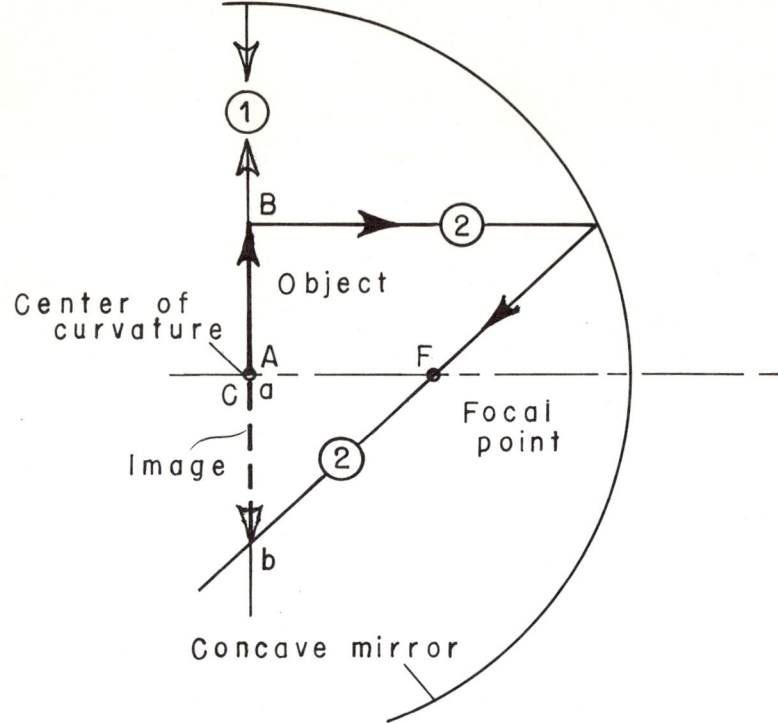

FIGURE 34. Position of the reflected image of an object located at the center of curvature of a concave mirror.

those for the convex mirror, where the image is always virtual. To study the several situations, we place the object *AB* at various positions before the mirror and diagram the rays. First, a ray from *B* is drawn through the center of curvature of the mirror, for this ray is reflected on itself. Second, a ray from *B* is drawn parallel to the axis *AC*, for this ray is reflected down through the focal point *F*. In Figure 32, where the object is at a considerable distance in front of the mirror, the image is *real*; it is *inverted* and smaller than the object. It is real because the rays from the object actually converge to an image. A photographic film at that position would be exposed by the light in the image.

Consider now the following two cases:

    1. The object is placed at the center of curvature of the concave surface (Fig. 34). The two rays chosen then intersect at the image point *b*, directly below the object point *B*. The image

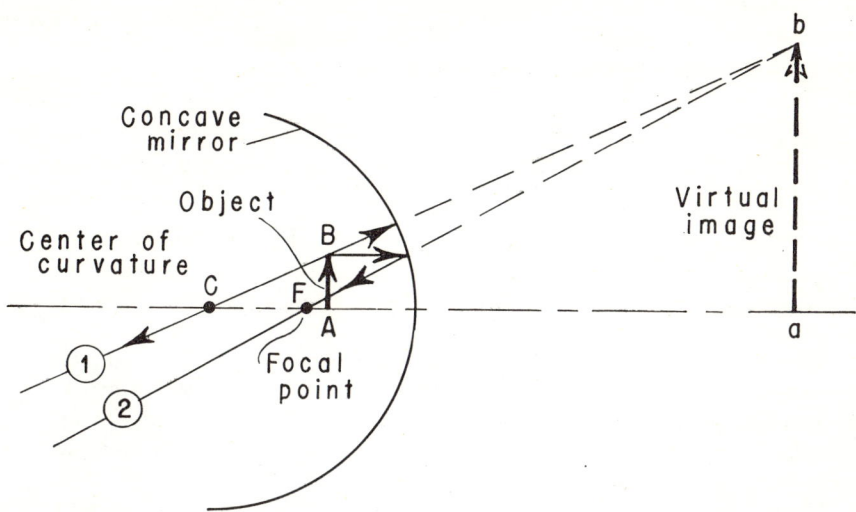

FIGURE 35. Position of the reflected image of an object located inside the focal distance of a concave mirror.

$ab$ is inverted, real, and of the same size as the object $AB$. This result follows also from the image equation, for if we substitute $u = R$, then

$$\frac{1}{v} - \frac{1}{R} = \frac{2}{-R} \quad \text{or} \quad \frac{1}{v} = \frac{1}{-R} \quad \text{or} \quad v = -R,$$

which means that the image is on the left side of the mirror surface at a distance $R$, which is the distance to the center of curvature. Also, the magnification $I/O$ ($= v/u$) becomes $-R/R = -1$; hence $I = -O$. The minus sign is taken to indicate an inverted image.

2. The object is placed inside the focal length of the mirror, as illustrated in Figure 35. The two selected rays upon reflection diverge *as if* from a point $b$ behind the mirror. Thus the image $ab$ of the object $AB$ inside the focal length of the mirror is an erect, magnified virtual image. This result also follows from the image equation. In general this gives

$$\frac{1}{v} = \frac{1}{u} - \frac{1}{f}.$$

When the object distance $u$ is smaller than $f$, $1/u$ is greater than $1/f$, and then the image distance $v$ is plus, which means the image is on the right side of the mirror and must thus be virtual. As the object distance $u$ becomes more nearly equal to the focal distance $f$, $1/v$ approaches zero, and the image distance $v$ approaches infinity. The magnification $I/O$ is plus and generally much greater than one.

*Examples:* In ophthalmology, the images reflected from the cornea have already been mentioned. However, each of the other three surfaces of the dioptric system of the eye also forms images by reflection. The posterior surface of the crystalline lens acts as a concave mirror, the other three as convex mirrors. The images from these surfaces are called *catoptric* images, or *Purkinjé* reflexes of the eye. Under suitable experimental arrangements all these images can be observed and photographed. There are, of course, more than just four catoptric images because of multiple reflections from the various surfaces of the eye. Actually, seven such images are usually described, but the brightness of the last three is very considerably less than that of the first four.

Spectacle lenses with their convex and concave surfaces are notorious for causing undesirable reflected images. Even the light from the bright image formed by reflection from the cornea is reflected by the posterior surface of the spectacle lens, and this image may often be seen in the field of vision.

The head mirror used by the otolaryngologist is a concave mirror used for concentrating light.

Other applications are found in the ordinary concave hand mirrors, in the Schmidt telescopic system, in reflecting microscopic objectives, in projector mirrors, and in searchlights. In the last the principle of image and object reversibility is used. A light source of high intensity is placed at the focal point of the mirror; hence a powerful beam of parallel-ray light (image at infinity) is projected into space.

Actually, convex and concave mirrors do not produce entirely stigmatic images. The point of the intersection of the reflected rays of adjacent incident rays varies for different parts of the mirror, as illustrated in Figure 36. The mirror is then said to exhibit *spherical aberration*. The farther the ray is from the axis, the greater is the deviation of the reflected ray from the paraxial focal point of the mirror. The surface formed by the loci of the adjacent points of intersection of the reflected rays is called the *caustic* surface. In any plane it is a caustic curve. The caustic curve or surface is said to be the envelope of the reflected rays.

It should be clear from this discussion that the object-image distance relationships for the concave and convex mirrors as derived in the preceding pages hold only for rays that are near the axis—the so-called *paraxial* rays. The usefulness of convex and concave mirrors is thus re-

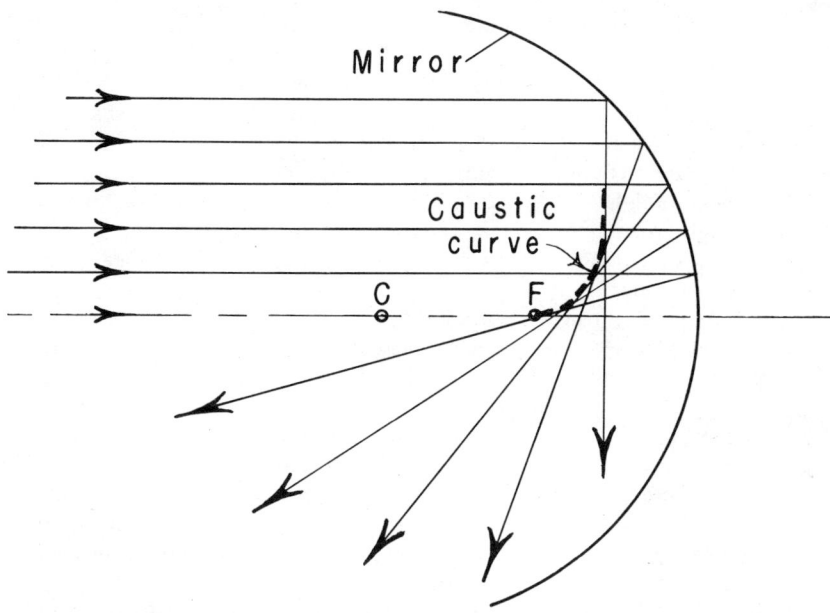

FIGURE 36. Caustic curve produced by the totality of the reflection of many rays from a concave mirror.

stricted, unless corrections are made either in the curvatures of the mirrors themselves or by optical correction plates. For certain types of searchlights or similar instruments the curvature is changed. It can be shown that spherical aberration is eliminated from a concave mirror by making the surface parabolic in shape; when this is done all reflected rays from a distant object intersect at the same point. The construction of accurate parabolic mirrors often is difficult and expensive. Correction plates are used in the Schmidt camera, and their equivalents are used in reflection objectives in microscopes. One of the advantages of a mirror over a lens is the absence of chromatic aberration.

## EXERCISES

1. A man is 6 feet tall. How long should a mirror be when placed 4 feet in front of him so that he obtains a full-length view of his reflected image? Diagram the rays.
2. Show by diagram how to find the image of the target on the arm of the haploscope, seen by the eye (see Fig. 170).
3. Show by diagram why the images of targets of a haploscope or amblyoscope, when turned from "far" to "near" vision, do not fall on the frontoparallel plane.

4. Two plane mirrors are hinged so that there is a 90° angle between the faces. Diagram the positions of the "three" images. What can be said about the position of the "third" image relative to the line of intersection of the two mirrors? Show with a compass that the three images and the object point itself all lie on a circle with center at the intersection of the mirrors.
5. A beam of light is reflected by a plane mirror. If the mirror is turned through an angle $\alpha$, show by diagram that the reflected beam will be rotated by $2\alpha$. This is the principle of the optical lever for measuring small rotations.
6. An illuminated ring 4 cm in diameter is placed at a distance of 10 cm in front of the cornea (a convex mirror), the radius of which is 8 mm. Where is the image of the ring and what is its size?
7. The apparent position of the pupil of the eye is near the position of the corneal reflex. Assuming an apparent pupil size of 3 mm, how large a circular object would be required at 50 cm from the eye to cause a corneal reflex as large as the size of the apparent pupil?
8. By an ophthalmometer, the size of the corneal reflex is measured to be 0.25 cm for a target of 20-cm diameter located 30 cm from the eye. What is the radius of curvature of the cornea?
9. An object 10 mm high is 100 mm in front of a concave mirror whose radius of curvature is 50 mm. Where is the image and what is its size?
10. Suppose an individual wants to view his face as reflected from an ordinary hand magnifying (concave) mirror. Let the mirror be 20 cm from the face. He wishes to have an image about 40 cm from the mirror. What should be the radius of curvature and the focal length of the mirror? What would be the magnification $(I/O)$ of the image?
11. In problem 10, if the observer were a presbyope with no refractive error, and the mirror were still 20 cm from the face, what should the radius of curvature be for clearest image?
12. The virtual image formed by the cornea lies 4 mm behind the cornea. Before the eye at a distance of 16 mm is a spectacle lens whose concave rear surface has a radius of 8 cm. What is the position of the image of the corneal reflex formed by the posterior lens surface? Could this image be seen clearly?
13. An object 10 cm high is placed 100 cm in front of a concave mirror whose radius of curvature is 50 cm. (a) Where is the image and what is its height? (b) If the object is moved 10 cm closer to the mirror, how far does the image move and how does its height change?

Chapter III

# REFRACTION OF LIGHT

## 1. REFRACTION OF LIGHT

It is a common experience to see that light changes its direction on passing from one optical medium to another. The position of a fish seen in a pool is not in the direction it appears to be. An inclined rod (a canoe paddle), part of which is in the air and part in the water, appears bent at the surface. When the direction of propagation of a ray of light is changed at the interface between two substances, the ray is said to be *refracted*.

The phenomenon of refraction is due to the fact that the velocity of light is less in a substance than it is in free space—that is, in a vacuum. Actual experiments have shown that the velocity of light is indeed less in water than it is in air. The velocity of light is different in various transparent substances and, in a very rough way, decreases as the density of the substance increases. The law of the refraction of light for the passing of a ray from one substance to another can be derived from the principles of the wave theory of light.

## 2. SNELL'S LAW

Suppose a beam of light from a distant source, moving through one medium, strikes the interface with a second medium at an angle of incidence, $i$, with the normal to the surface (Fig. 37). Suppose the velocity of light in the second medium, $V_2$, is less than that in the first medium, $V_1$, as would be the case if the second medium were water and the first were air. We would like to find a relationship that specifies the change in direction of the beam at the interface—that is, the relationship between the angle of incidence ($\angle i$), the angle of refraction ($\angle r$), and the two velocities of light.

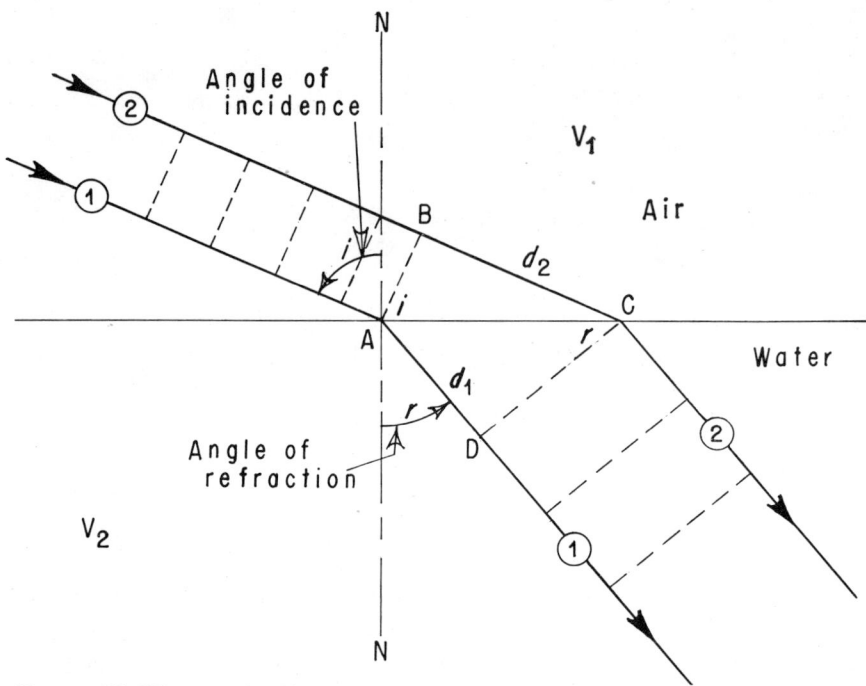

FIGURE 37. Diagram for deriving the law of refraction of light at the interface between two different optical media.

Since the incident light is from a distant source, the wave fronts are plane and perpendicular to the direction of propagation. At the instant ray 1 strikes the interface at $A$, the point $B$ on the same wave front on ray 2 must still travel a distance $d_2 = BC$ before striking the interface. The time $t$ necessary for ray 2 to travel this distance will be $t = d_2/V_1$ (distance divided by velocity). During this time $t$, however, ray 1 will have traveled a distance $d_1$ in the second medium in some direction to some position $D$. This distance would be $d_1 = V_2 t$. The points $D$ and $C$ must be in phase; hence they will lie on the same wave front in the second medium. The direction, then, of the refracted beam would be at right angles to this wave front. The ratio of the two distances $d_2 \, (=V_1 t)$ and $d_1$ would be

$$d_2/d_1 = V_1/V_2.$$

To make this ratio useful, consider the two right triangles $ABC$ and $ADC$. The distance $AC$ is common to both; in fact, it is the hypotenuse of each. The angle $BAC$ also is identical to $\angle i$, the angle of incidence, and the angle $ACD$ also is identical to $\angle r$, the angle of refraction. We wish to find the relationship between the distances $d_1$ and $d_2$ in terms of the angles of incidence and refraction.

*Note:* For any series of right triangles with the same included angle, $\angle \alpha$ (Fig. 38), the ratio of the opposite side $a$ to the hypotenuse $h$ of any triangle is the same constant number because these right-angle triangles are similar and the sides therefore are proportional. The ratio (a number) $a/h$, for a constant angle $\alpha$, has a special designation in trigonometry—namely, the sine (written *sin*), so that

$$\sin \alpha = a/h, \text{ whence } a = h \sin \alpha.$$

When the angle $\alpha$ is specified (say, so many arc degrees), we can use appropriate tables to find the value of the sine of that angle to any desired accuracy.

For the two triangles under discussion (Fig. 37) we can write

$$d_2 = AC \sin i \text{ and } d_1 = AC \sin r.$$

Dividing the distance $d_2$ by distance $d_1$, and substituting the equivalent ratio above, we obtain

$$\frac{d_2}{d_1} = \frac{\sin i}{\sin r} = \frac{V_1}{V_2} = n. \quad \ldots (9)$$

The number $n$ is the ratio of the velocities of light in the two media—that is, $V_1/V_2$. This number is called the *relative index of refraction* between these two substances. For air and water it is 1.333, a number determined only by experiment.

The *absolute* index of refraction of a substance is the ratio of the velocity of light, $V_0$, in a free space (a vacuum) to the velocity in the substance. In the present example, the absolute index of refraction of the first medium would be $n_1 = V_0/V_1$, and that of the second medium would be $n_2 = V_0/V_2$. Dividing the second of these by the first, we obtain

$$n_2/n_1 = V_1/V_2 = n.$$

Equation 9 can then be written

$$\sin i / \sin r = n_2/n_1.$$

More generally, if $\theta_1$ and $\theta_2$ replace $i$ and $r$, respectively, then

$$\underbrace{n_1 \sin \theta_1}_{\text{first medium}} = \underbrace{n_2 \sin \theta_2}_{\text{second medium}} \quad \ldots (10)$$

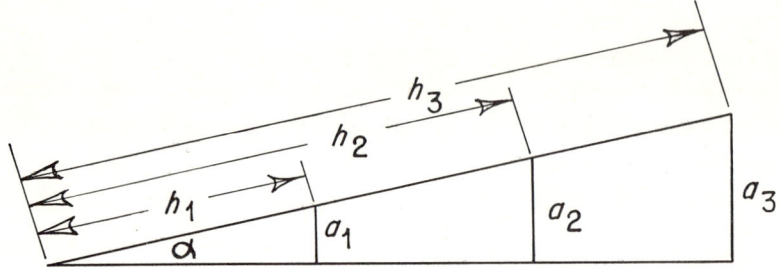

FIGURE 38. Similar triangles used to illustrate the sine function of a right triangle for the included angle $\alpha$.

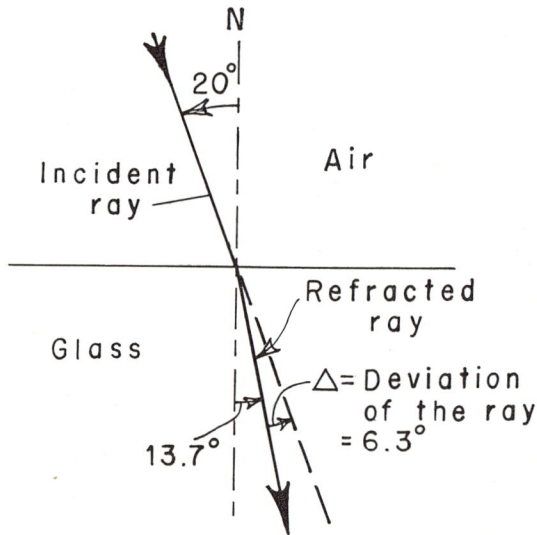

FIGURE 39. Refraction of a ray passing from air into glass.

This is the fundamental law for the refraction of light and is known as *Snell's law*, after the man who discovered it. This law, which accurately describes quantitatively the refraction of light by transparent substances, is the most useful relationship in optics. Upon it depends all lens design. It should be kept in mind that this relationship holds also when the light ray travels in the reverse direction.

The greater the difference between the indices of refraction—that is, the greater the relative index between the two media—

the greater is the deviation of the refracted ray. For a given relative index of refraction, the deviation of the refracted ray is not proportional to the angle of incidence but is proportional to the sine of this angle. Equal steps of angle are not proportional to equal steps of the sine of the angle (except approximately for very small angles). Because the sine of an angle is the ratio of the opposite side of a right triangle to the hypotenuse of the triangle (Fig. 38), as the angle $\alpha$ becomes greater and greater, the length of the opposite side $a$ approaches the length of the hypotenuse $h$; so the sine of the angle $\alpha$ never can be greater than one. Actual values for the sine are shown below.

| Angle | 0° | 10° | 20° | 30° | 40° | 45° | 50° | 60° | 70° | 80° | 90° |
|---|---|---|---|---|---|---|---|---|---|---|---|
| Sine | 0.000 | 0.173 | 0.342 | 0.500 | 0.643 | 0.707 | 0.766 | 0.866 | 0.940 | 0.985 | 1.00 |

*Example 1:* As an example (Fig. 39), suppose a ray of light passes from air, whose index of refraction is $n_1 = 1.00$, to another medium—say, glass, whose index of refraction is $n_2 = 1.50$. The angle of incidence $\theta_1$ is given as 20°. The angle of refraction $\theta_2$ then can be calculated from equation 10. We have, then, $\sin 20° = 1.5 \sin \theta_2$, from which $\sin \theta_2 = \sin 20°/1.5$. Using tables, we find $\sin 20° = 0.342$. Thus, $\sin \theta_2 = 0.342/1.5 = 0.228$. Again using tables, we find for the angle of refraction $\theta_2 = 13.18°$. The deviation of the ray from its original direction is then $\Delta = \theta_1 - \theta_2 = 20° - 13.18° = 6.82°$.

*Rule:* Whenever a ray of light passes from a substance into another substance having a *higher* index of refraction, the angle of refraction always is *smaller* than the angle of incidence; that is, the ray is deviated *toward* the normal to the interface.

*Example 2:* Clearly, if we reverse the direction of the light ray in Figure 39, the angle of incidence would be 13.7°; the angle of refraction would be 20°.

*Rule:* Whenever a ray of light passes from a substance into another substance having a *smaller* index of refraction, the angle of refraction is *greater* than the angle of incidence; that is, the ray is deviated *away from* the normal to the interface.

## 3. CRITICAL ANGLE OF REFRACTION

In the second example given above (glass to air), the index of refraction of the second substance is smaller than that of the first; therefore the angle of refraction is larger than the angle of incidence. There is, therefore, a *critical angle* of incidence, $\theta_c$, for which the angle of refraction for the refracted ray will be 90°; that is, the

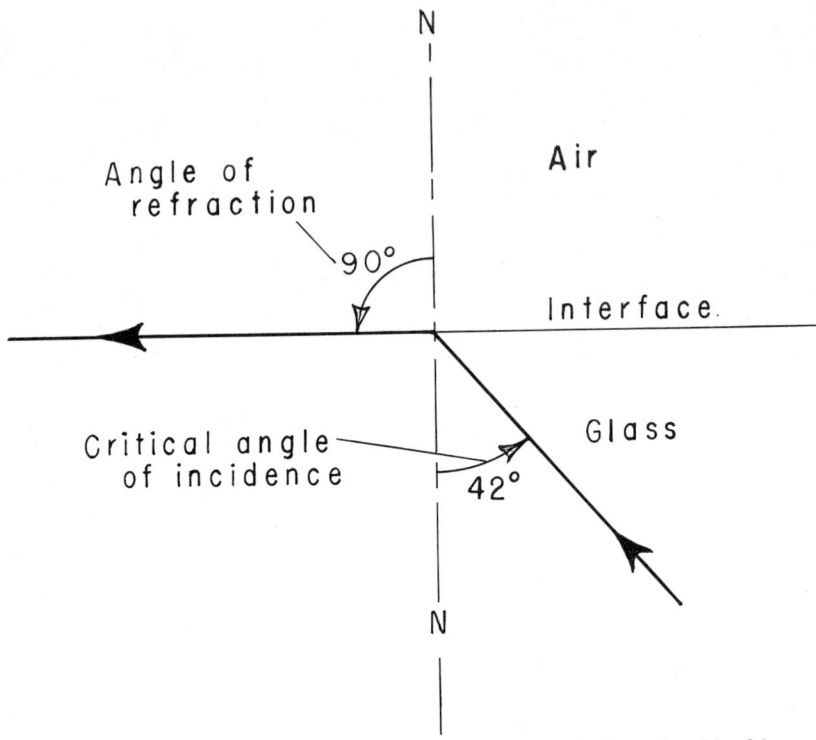

FIGURE 40. Phenomenon of total reflection at the critical angle of incidence.

refracted ray will coincide with the surface of the interface (Fig. 40). We can calculate the magnitude of the critical angle for a ray passing from glass to air from Snell's law because from equation 10

$$1.5 \sin \theta_c = 1.0 \sin 90° = 1.$$

Thus,
$$\sin \theta_c = 1/1.5 = 0.67,$$
whence the critical angle (using tables)
$$\theta_c = 42°.$$

The index of refraction of a liquid or solid can be measured by the Abbé refractometer, which depends upon determination of the critical angle. This instrument is calibrated to read as follows: $n = 1/\sin \theta_c$.

For angles of incidence greater than this critical angle (Fig. 41), the incident ray will be wholly *reflected* within the first medium.

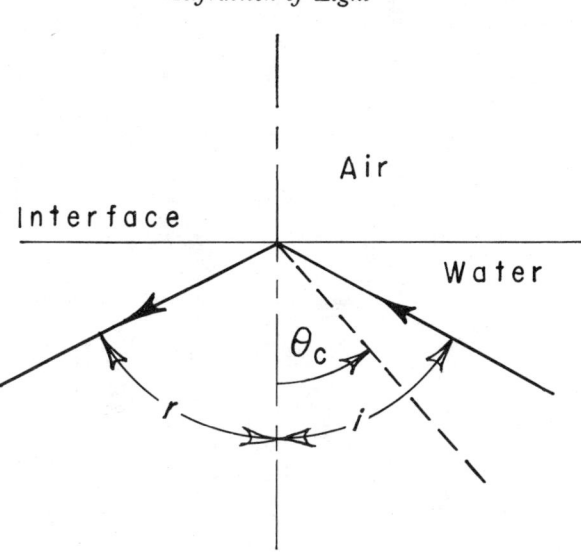

FIGURE 41. Phenomenon of total reflection when the angle of incidence is greater than the critical angle.

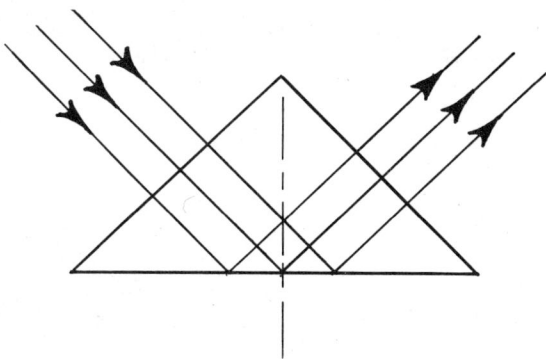

FIGURE 42. A right-angle prism reflecting light as from a mirror, in accord with the principle of total reflection.

The important application of this phenomenon is the use of prisms in certain optical instruments to replace mirrors (Fig. 42). Whenever the angle of incidence of the rays at the *internal* surface of the prism is greater than 42°, the rays are totally reflected. At angles less than 42°, some of the light of the rays is reflected and some is refracted. Total reflecting prisms are used in many optical instruments such as binoculars and the ophthalmoscope.

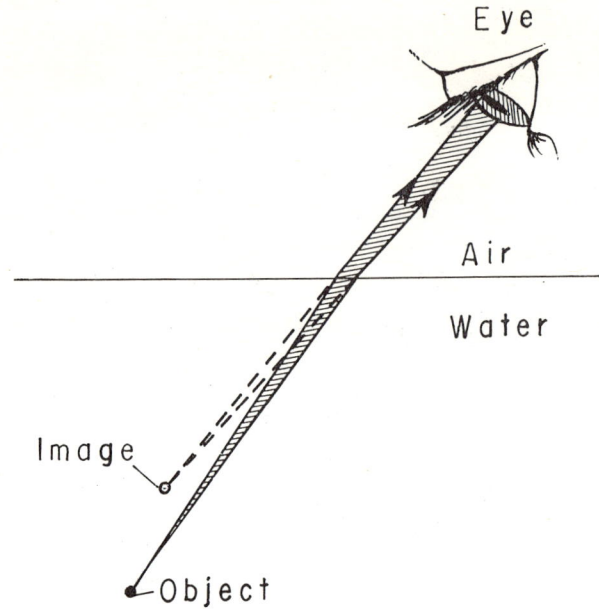

FIGURE 43. The particular pencil of light from a submerged object point reaching the eye, and the corresponding position of the image of the object.

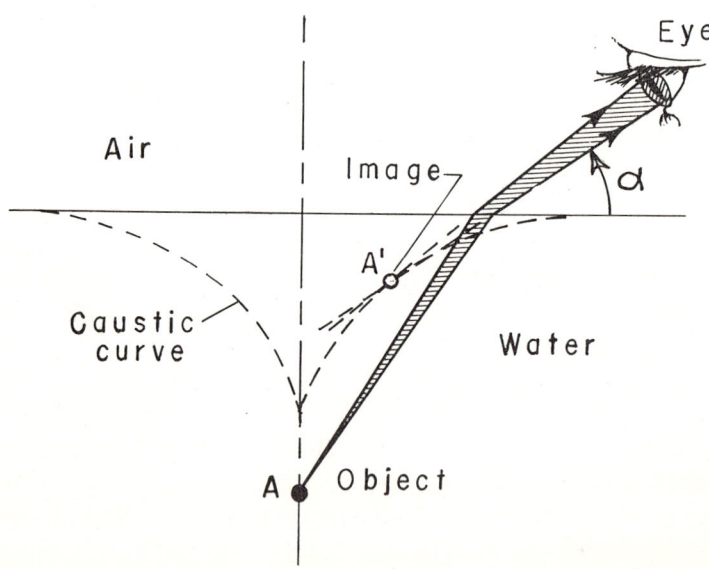

FIGURE 44. Curve for the loci of images formed by refraction, for different inclinations of the direction of observation.

## 4. IMAGE FORMATION BY REFRACTION AT A PLANE SURFACE

When we observe an object submerged in water or in other media we see a definite image of that object, although the image does not generally coincide with the object itself (Fig. 43). The position of the image is determined by those rays included within a small cone from the object as limited by the size of the pupil. The rays, then, are nearly adjacent. We can predict where this image will be on the basis of the law of refraction of light. Experiment and calculation both show, however, that the position of this image, again determined by a small cone, varies with the angle at which the image is being viewed by the eye. The loci of the images of successive rays are illustrated in an exaggerated manner in Figure 44. As the direction of observation moves from the normal to one increasingly oblique, the image moves along a curve, which is called a *caustic* curve. This curve, of course, is only a plane section of a caustic surface.

We shall be concerned here only with the position of the image under observation nearly normal to the surface of the interface. To determine this position we need consider only two rays from the submerged object point (Fig. 45). We select first the ray (1) that passes perpendicularly to the surface, for this ray will be undeviated (since the angles of incidence and refraction are both zero). The second ray (2) from $A$ makes an angle of incidence, $\theta_1$, to the normal to the surface. This angle is shown large in order to make the diagram clear. This ray is incident to the surface at some point $B$ and is refracted so as to emerge at an angle $\theta_2$ to the normal. The ray appears to have come from the direction $A'B$, and the intersection of this produced line and the perpendicular line at point $A'$ is the position of the image.

Let the distance of the object $A$ below the surface be $t$ and that of the image be $t'$. To find the relationship between these, we can make use of the two right triangles $APB$ and $A'PB$. From the simple trigonometric relations shown in the supplementary drawing (Fig. 46), we can write for the distance $PB$, which is common to both the triangles, the following:

$$PB = t \tan \theta_1 = t' \tan \theta_2,$$

whence

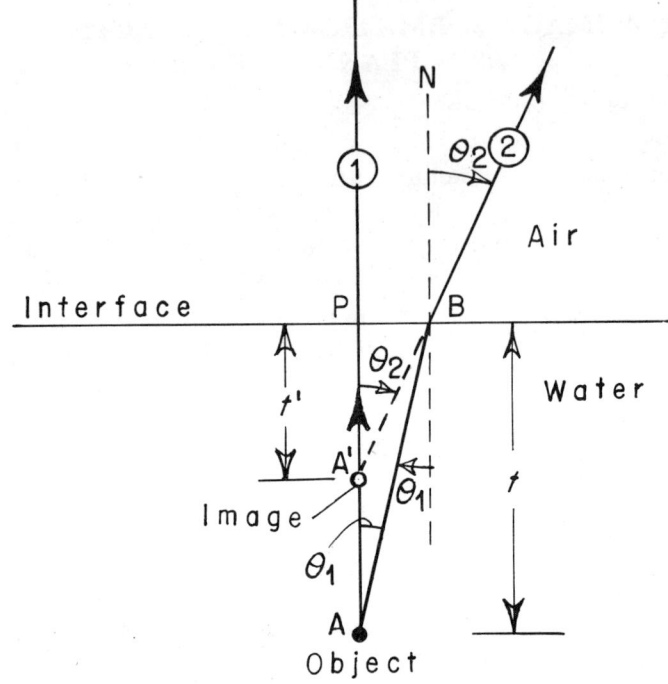

FIGURE 45. Diagram used in deriving the relationship between the positions of the object and of the image formed by refraction from a plane interface.

$$t' = t \tan \theta_1 / \tan \theta_2 = t[\sin \theta_1/\cos \theta_1]/[\sin \theta_2/\cos \theta_2]$$

or

$$t' = t\,[\sin \theta_1/\sin \theta_2]\,[\cos \theta_2/\cos \theta_1]\,. \qquad \ldots (11)$$

However, from Snell's law (equation 10), $\sin \theta_1/\sin \theta_2 = n_2/n_1$. Substituting this into equation 11, we obtain

$$t' = t\frac{n_2}{n_1}\frac{\cos \theta_2}{\cos \theta_1}.$$

This relationship shows that the depth of the image varies with the angle of incidence. However, with nearly normal observation, $\theta_1$ and $\theta_2$ are both very small, so both $\cos \theta_1$ and $\cos \theta_2$ are near one* and their ratio is even closer to one. Therefore, for near normal observation, the apparent depth of the image of the submerged object is given by

---

*This follows because (see Fig. 46) as the angle becomes smaller and smaller, $b$, the adjacent side of the triangle, becomes more nearly equal to $h$, and thus their **ratio approaches one.**

# Refraction of Light

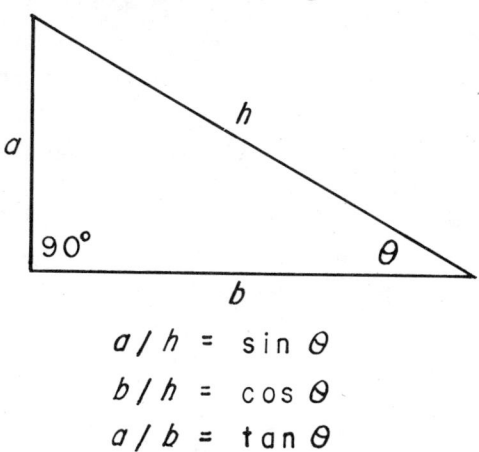

FIGURE 46. The trigonometric functions of the included angle of a right triangle.

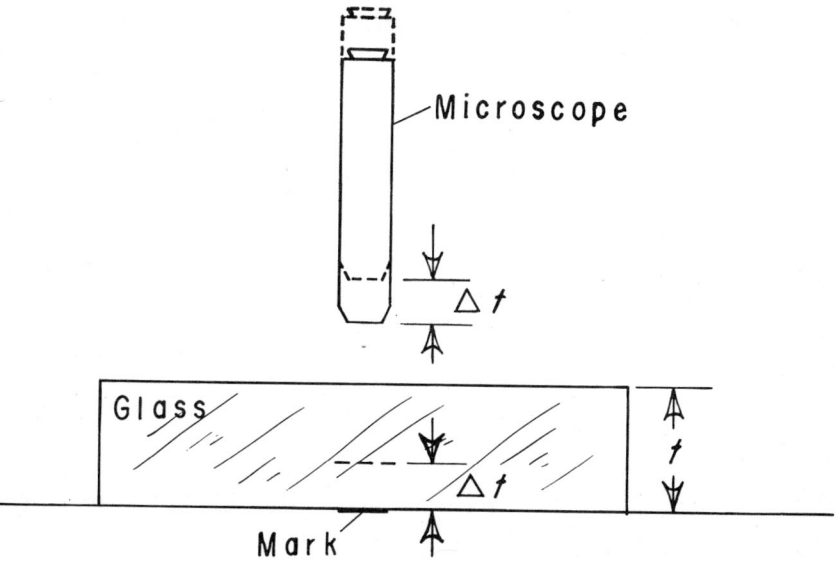

FIGURE 47. A method for determining the index of refraction of a glass plate.

$$t' = t\,(n_2/n_1). \qquad \ldots (12)$$

The difference in depth ($\Delta t$) of image and object would be

$$\Delta t = t - t' = t - t n_2/n_1 = t\,(n_1 - n_2)/n_1. \qquad \ldots (13)$$

In this relationship, of course, the light emanates from the object $A$ (Fig. 45) within the water and travels from a medium of greater index ($n_1$) out into air of smaller index ($n_2$).

The index of refraction of a given substance can be determined with fair precision by measurement of this difference—for example, by a vertically traveling microscope. Suppose we desire to learn the index of a glass plate (which has polished parallel sides). A piece of thin paper bearing a small mark is placed upon the tabletop, and the microscope is moved up and down until the image of the mark is exactly in focus (Fig. 47). Then the plate of glass is placed upon the paper, and the microscope is raised a distance $\Delta t$ in order to focus again on the image of the mark as now seen through the glass plate. If the thickness of the glass is $t$, then from equation 13, $n_1/n_2 = t/(t - \Delta t)$. Since $n_2$ is the index of refraction of air and is equal to 1.00, the ratio $n_1/n_2$ is the index of refraction of the glass used in the plate. (Although the index of refraction of a given sample of glass can be determined in this way, more precise means are used which are mostly dependent upon measurement of the critical angle.)

## 5. VARIATION OF INDEX OF REFRACTION WITH WAVELENGTH

In free space—that is, a vacuum—the velocity of light is the same for light of all wavelengths. This is not true in refracting media, for there the index of refraction differs with the different wavelengths (being greatest for blue, least for red) and with the different media. When we speak, loosely, of "the index of refraction" of a given type of glass, for example, we usually refer to the index of refraction ($n_D$) for the D Fraunhofer spectral line (589 m$\mu$)—the yellow sodium line—of the spectrum. Examples of the $n_D$ indices of refraction for certain substances are as follows:

```
Air (standard conditions).....................1.00029
Water ........................................1.333
Spectacle crown glass.........................1.523
Aqueous and vitreous humors...................1.336
Cornea .......................................1.376
```

The degree to which a given glass will separate white light into its component colors can be quantified when the indices of refraction are known for the C (red), D (yellow), and F (blue) lines of the Fraunhofer spectrum. The magnitude of the *relative dispersion* of the glass is specified by a ratio called $\nu$ (nu), which is defined by $\nu = (n_D - 1)/(n_F - n_C)$. The larger the value $\nu$, the smaller is the dispersive power of the glass. Selected examples of the indices of refraction and of the corresponding $\nu$ values of certain optical substances are as follows:

| Substance | C (658 mµ) | D (589 mµ) | F (486 mµ) | $\nu$ |
|---|---|---|---|---|
| Beryllium fluoride | 1.274 | 1.275 | 1.276 | 107 |
| Spectacle crown glass | 1.520 | 1.523 | 1.529 | 60 |
| Barium crown glass | 1.565 | 1.568 | 1.576 | 49 |
| Light flint glass | 1.576 | 1.580 | 1.590 | 42 |
| Medium flint glass | 1.611 | 1.616 | 1.628 | 38 |
| Densest flint glass | 1.879 | 1.890 | 1.919 | 20 |
| Polymethylmethacrylate (plastic) | 1.489 | 1.491 | 1.497 | 61 |
| Fused quartz glass | 1.456 | 1.458 | 1.463 | 65 |
| EK-448 glass | 1.798 | 1.880 | 1.896 | 41 |
| Water (20° C) | 1.331 | 1.333 | 1.337 | 56 |

It is this relative dispersion of glass that produces the color fringes about objects observed through prismatic spectacle lenses and through certain types of bifocals. Color-corrected lenses employ two or more types of glass having different $\nu$ values.

## 6. PLANE PARALLELS OF GLASS

The ray of light passing obliquely through a sheet of glass with parallel sides emerges from the glass at the same angle at which it enters (Fig. 48). This should be clear from an inspection of the figure. However, the ray will be displaced a lateral distance, $s$,

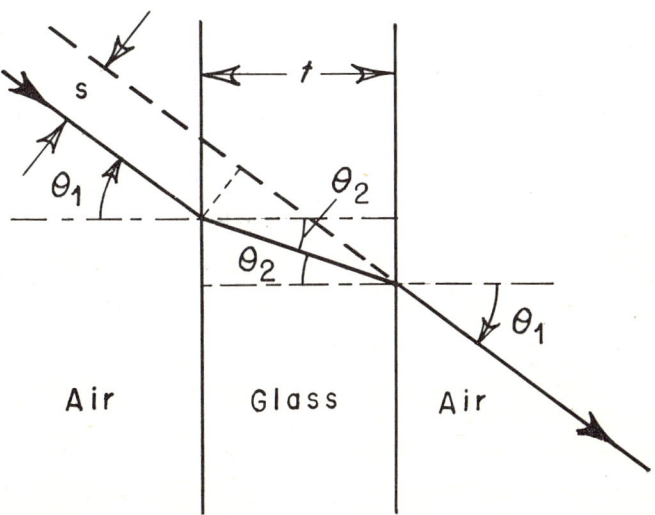

FIGURE 48. Displacement of a ray passing through a plane-parallel glass at an inclined angle.

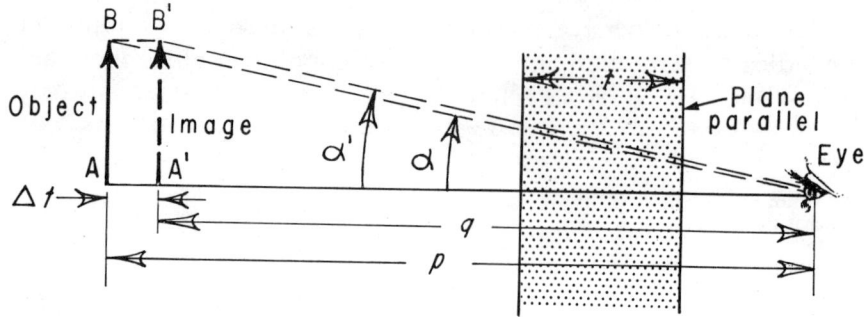

FIGURE 49. Small angular magnification of the image of a near object viewed through a plane-parallel glass.

which depends upon the thickness of the plate, upon the index of refraction of the glass, and upon the angle of incidence. If an object *near* an obliquely placed plane parallel is observed, the angular displacement introduced by $s$ may be important—for instance, in the haploscope. If the object is at a considerable distance from the eye, the angular displacement due to the plane parallel is negligible for usual purposes.

The image of an object viewed through a plane parallel of glass is displaced toward the eye by the amount calculated in section 4 of this chapter. This displacement introduces a slight angular magnification of the image, since it is nearer the eye than is the object (Fig. 49). Prior to the introduction of the plane parallel before the eye, the angular size of the object $AB$ is $\alpha$. With the plane parallel, the image becomes $A'B'$, being displaced a distance $\Delta t$ nearer the eye, and the angular size of this image is denoted by $\alpha'$. If the distance of the object from the eye is $p$, then the tangent $\alpha = AB/p$. The distance of the image from the eye is $p - \Delta t$, and the angular size is now given by tangent $\alpha' = A'B'/(p - \Delta t)$. From equation 13, $\Delta t = t(n_1 - n_2)/n_1$, where $t$ is the thickness of the plate. Since $A'B' = AB$, the angular magnification, $A$, of the object produced by the plane parallel is $A = \tan \alpha'/\tan \alpha = 1/[1 - t(n-1)/pn]$. In this $n_1 = n$—the index of refraction of the glass—and $n_2 = 1$.

It will be clear that if the object distance $p$ is large, the fraction $t/p$ approaches zero, and the magnification becomes unity—there is no increase in angular size. Thus, the angular magnification of objects viewed through plane parallels is important only when relatively near objects are observed. Since the magnifying effect is at any rate small, the percentage of angular magnification would be as follows:

$$\% \text{ angular mag.} = 100 \cdot [t/p]\,[(n-1)/n].$$

*Example:* Suppose one views an object at 33 cm through a pane of glass 5 mm thick. The percentage of angular magnification then is

$$\% \text{ angular mag.} = \left[\frac{.5}{33}\right]\left[\frac{1.5-1}{1.5}\right] \cdot 100 = 0.5\%.$$

A difference of 0.5 per cent in magnification between the two eyes can be discriminated in a specially designed instrument by the person with normal visual acuity.

Chapter IV

# OPHTHALMIC PRISMS

## 1. PRISMATIC DEVIATION

THE optical phenomena associated with ophthalmic prisms can be considered now.

Any optical substance bounded by surfaces that form a wedge constitutes a prism. These surfaces may be planes or portions of cylinders or of spheres. The edge at which the two surfaces would intersect is referred to as the *apex*. The angle between the faces of the prism at the edge is called the *apex angle*. The *base* of the prism refers to the thicker portion opposite the edge (Fig. 50). A plane through the prism perpendicular to the edge is defined as the *principal section* of the prism and as such is said to be the *base-apex meridian*.

A pencil of light passing through the prism is deviated from its original direction, and at the same time the light is dispersed into its component spectral colors. Here we shall be interested primarily in the deviation and shall assume, for the time being, at least, that the incident light is monochromatic.

It is important to note (Fig. 51) that when the eye observes an object-point source of light through a prism, the diameter of the effective cone of light is limited by the size of the pupil; hence that portion of the prism through which the cone passes is relatively small. Furthermore, the cones of light from other object points in the field of view must pass through different portions of the prism. This fact results in slightly different prismatic deviations of the images of different points in the visual field.

The angular deviation of rays passing through a prism varies somewhat according to how the prism is oriented (Fig. 50).

First, consider the simple case in which a prism with plane surfaces is placed before the eye so that the front surface is perpen-

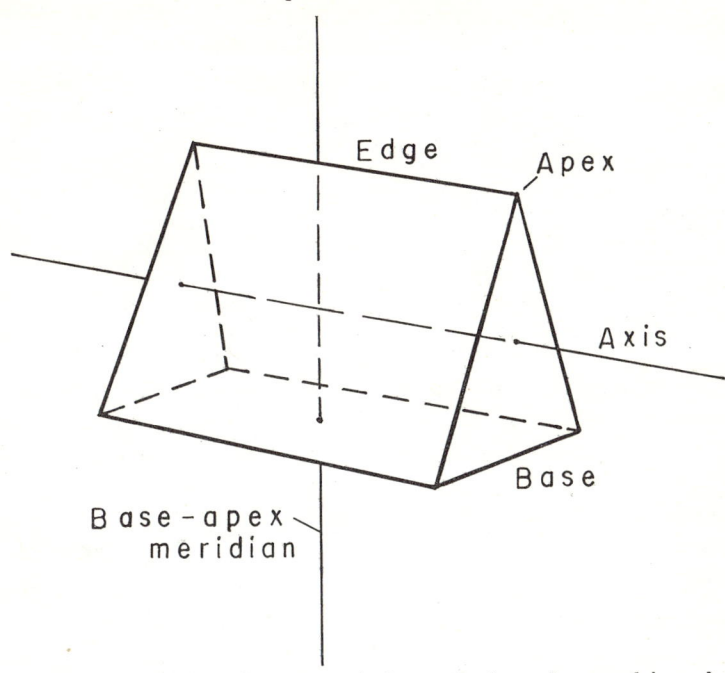

FIGURE 50. Names of certain geometric features of a prism used in ophthalmic optics.

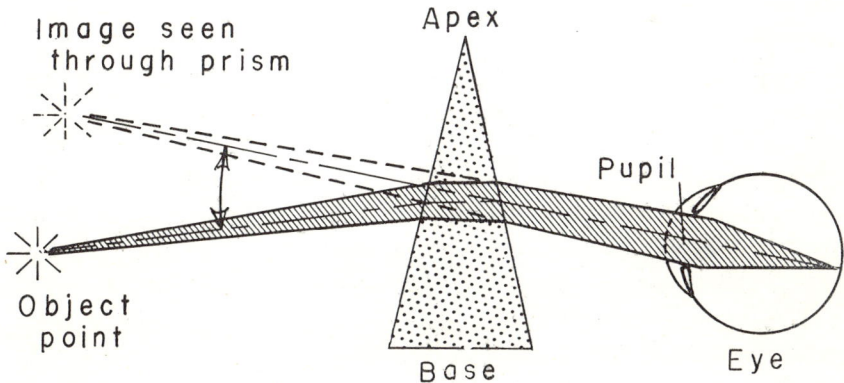

FIGURE 51. Course of a pencil of light from a point object through a prism to the pupil of the eye, showing the restricted area of the prism actually used.

dicular to the direction of the object being observed (Fig. 52). The rays from a distant object will then fall perpendicular to the front face of the prism. Consequently, that central ray, on traversing

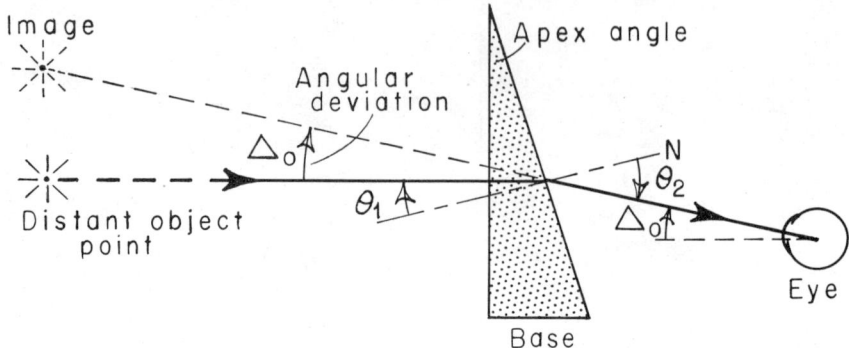

FIGURE 52. Diagram used in deriving the relationship between the prismatic deviation and dimensions of the prism.

the prism and reaching the eye, passes through the first surface undeviated but on passing through the second is deviated toward the base of the prism. The subject sees the image of the object, however, as deviated toward the apex of the prism. When the eye is fixating a given object point and a prism is placed before the eye, the eye will have to turn in a direction toward the apex in order to fixate the image of the object point. Thus, if one wishes to cause the eye to turn up, he must place the prism before it base-down.

We wish to find the relationship of the deviating power in terms of the apex angle of the prism. An inspection of Figure 52 shows that the angle of incidence of the ray at the second surface, $\theta_1$, also is equal to the apex angle, $A$. This follows, since the angle $\theta_1$ is measured between the normal to the first surface and the normal to the second surface at the point of incidence. The angle between the perpendiculars (normals) is equal to the included angle between the surfaces. Referring to Snell's law, we can write $\sin \theta_2 = n_1 \sin A$, since the index of air is taken as unity and $n_1$ is the index of refraction of the glass. The angular deviation, $\Delta_o$, caused by the prism would be

$$\Delta_o = \theta_2 - \theta_1 = \theta_2 - A.$$

Thus, $\Delta_o = \arcsin (n_1 \sin A) - A$, a relationship which reads as follows: The angular deviation caused by the prism is equal to the angle whose sine is equal to $(n_1 \sin A)$, minus the apex angle $A$.

## Ophthalmic Prisms

For many of the ophthalmic prisms used clinically, the prismatic deviations are small, and the apex angle $A$ therefore is small. For these small angles we may therefore use in the above relationship the angle itself, instead of the sine, without loss of precision. Then $\theta_2 = nA$, and the angular deviation is

$$\Delta_o = nA - A = A(n-1). \qquad \ldots (14)$$

Since the index of refraction of spectacle glass is about 1.5, then $(n - 1) = 0.5$, and we can write with sufficient accuracy that the prismatic deviation of a prism with a small apex angle is $\Delta_o = \frac{1}{2}A$, both $\Delta_o$ and $A$ here being in arc degrees. This is a useful approximate relationship to remember. Ophthalmic prisms were at one time specified by the apex angle, in arc degrees.

## 2. ORIENTATION OF PRISMS FOR MINIMAL DEVIATION

The preceding discussion is strictly accurate only when the prism is held before the eye so that the front face is perpendicular to the direction of the incident rays.

It can be shown experimentally and theoretically that if the prism is rotated about the prism axis from this position clockwise (Fig. 53), the prismatic deviation first decreases and then increases. The minimal deviation will occur when the ray traverses the interior of the prism symmetrically: The angle of incidence to the front surface is equal to the angle of refraction at the back surface, and the ray traversing the prism is perpendicular to a plane that bisects the apex angle. The relationship between the prismatic deviation and the angle of rotation of the prism is shown

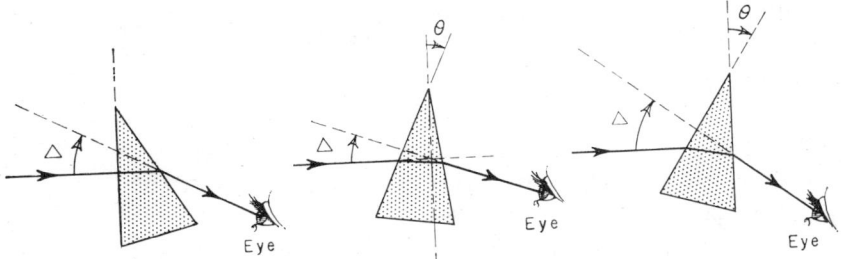

FIGURE 53. The prismatic deviation of a prism as affected by its orientation before the eye.

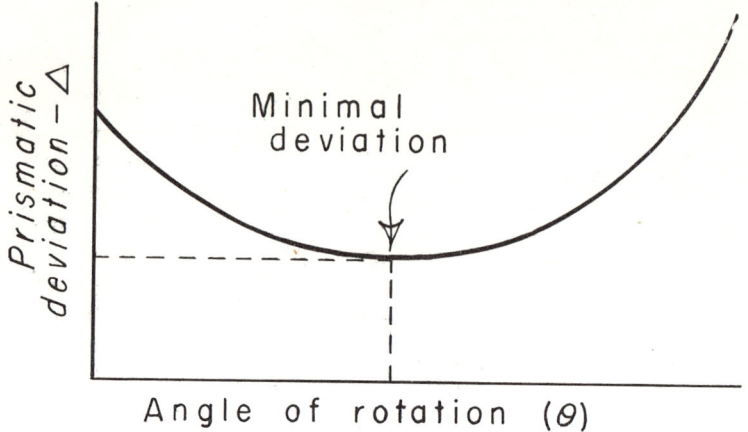

FIGURE 54. Graph of the relationship between the prismatic deviation of a prism and different rotated positions, showing especially the existence of a minimal deviation for a particular position.

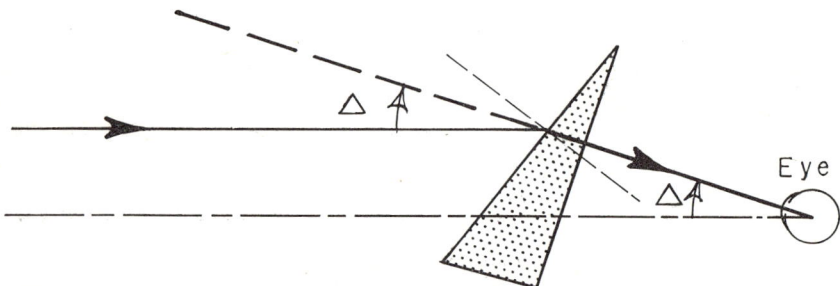

FIGURE 55. Prismatic deviation when the back face of the prism is at right angles to the visual axis.

schematically in Figure 54. Although the position at which the prism is held before the eye in measuring ocular deviations is relatively unimportant for prisms of low deviating power, it becomes important for the higher powers if precision in measurement is desired. The prismatic deviation stamped on the prism is correct only when the prism is oriented before the eye in a specific manner.

If the prism is oriented so that the back surface is perpendicular to the line of sight of the eye when one is looking through the prism (Fig. 55), the prismatic deviation will be the same as when

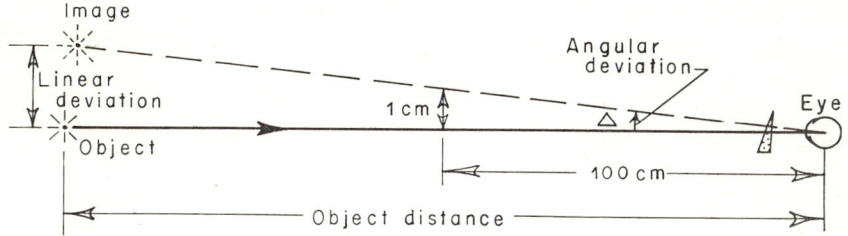

FIGURE 56. Diagram illustrating the unit of prismatic deviation—the prism diopter.

the prism is oriented so that the front surface is perpendicular to the principal incident ray. The sets of ophthalmic prisms used for measuring strabismus are usually calibrated corresponding to the deviation produced when they are held before the eye with the back surface normal to the visual axis of the eye.

## 3. THE MEASURE OF PRISMATIC DEVIATION

Although for physicists the specification of the prism angle and of the refractive characteristics of the glass used is sufficient, the clinician—being concerned only with the effect produced by the prism when held before the eye—wants the prism labeled according to this effect. From a rigorous point of view there is considerable confusion and inconsistency in the specification of prismatic deviation of ophthalmic prisms.

The arc degree would be a logical unit because it is so generally used to specify angles, but for the small prismatic deviations most frequently encountered this would perhaps result in small fractions, which apparently is objectionable. The prism diopter (originally proposed by Prentice[2] and usually designated by the exponent $\Delta$) has become by usage the most favored.

In fact, as prisms are used, the specification of the deviating power of a prism should measure the angular displacement of the image from the object relative to the center of rotation of the eye behind the prism. If this *angular* deviation *corresponds* to an angle which subtends 1 cm at a distance of 1 meter from the eye, that deviation produced by the prism is said to be 1 prism diopter ($1^\Delta$) (Fig. 56). Because the deviating powers of most prisms used clinically are small, the factor of the distance of the prism from the eye may be neglected.

In order that we may have a system for denoting prism power in which the deviation produced by one prism can be added to (or subtracted from) the deviation produced by another prism, the prism diopter as defined should be a *unit* of angular measure. The prismatic deviation of a given prism should therefore be expressed in multiples of this unit.* In spite of the definitions to the contrary found in some textbooks, this method of specifying the deviating power of prisms is the one actually being used clinically—although it is beset with some lack of precision.

The prism diopter is the unit of measure of an angle. Since the tangent of the angle corresponding to a deviation of 1 cm at a distance of 1 meter is $\tan \Delta = 1/100 = 0.01$, the prism diopter is one hundred times that tangent. The angle which has for its tangent 0.01 is 0.57 arc degree; hence, a prism diopter is equal to 0.57 arc degree. Conversely, an angular deviation of 1 arc degree is equal to 1.74 prism diopters. At the end of section 1, it was shown that the angular deviation of a prism (arc degrees) is approximately equal to one half the apex angle. Approximately, then, the angular deviation in prism diopters will be equal to the apex angle in arc degrees. It should also be remembered that the maximal apex angle a prism can have, when oriented so that the incident ray is perpendicular to the front surface, is about 41°; for prisms with a larger apex angle than this there would be total internal reflection of incident rays within the prism.

## 4. EFFECT OF OBJECT DISTANCE

When the object being viewed through a prism is relatively near the eye, the deviation produced by the prism, insofar as the eye is concerned, is somewhat less than that produced by the same prism when the object being observed is at a considerable distance. The decrease in prismatic deviation for near objects depends on the distance of the prism from the eye as well as on the object distance. Usually, prisms are marked according to the deviation

---

*The prism diopter for the specification of the deviating power of a prism as used here is more akin to the seldom used *centrad*, a unit that will not be discussed here.

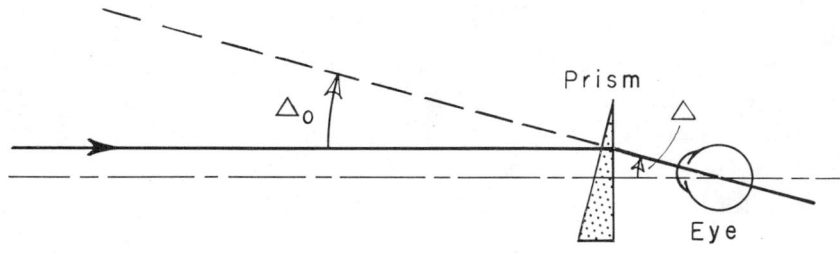

Distant Object

FIGURE 57. Prismatic deviation of a prism used with a very distant object.

caused by the prism for a very distant object. In this case (Figs. 52 and 57) the angle $\Delta$, which is due to the prism and through which the eye must turn in order to view the image of a very distant object, is identical to the actual angular optical deviation $\Delta_o$.

For a near object (Fig. 58), the angle $\Delta$ through which the eye must turn is smaller by a small angle $\epsilon$ (epsilon). This angle results because that chief ray from the object which, when deviated by the prism, will pass through the center of rotation of the eye must pass through the prism more toward the base of the prism. The total actual deviation $\Delta_o$ produced by the prism remains the same as before. Thus, $\Delta_o = \Delta + \epsilon$ (the external angle of a triangle is equal to the sum of the opposite interior angles). The actual effective prismatic deviation, $\Delta$, for the near object, would be $\Delta = \Delta_o - \epsilon$. This actual prismatic deviation can be determined with sufficient accuracy as follows: In Figure 58, let $p$ be the object distance from the eye and $h$ the distance of the prism from

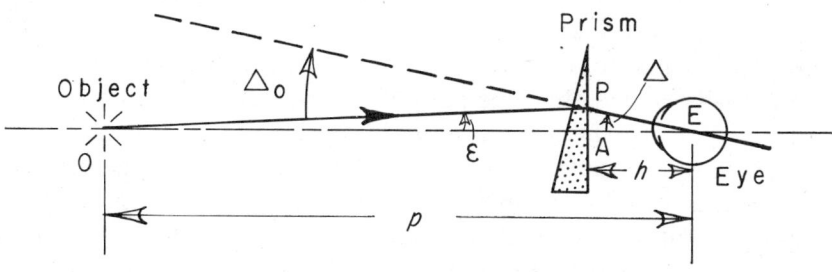

Near Object

FIGURE 58. Prismatic deviation reduced by nearness of object viewed through it.

the eye. The two triangles $OAP$ and $EAP$ have the side $AP$ in common. We can write therefore: $h \tan \Delta = (p - h) \tan \epsilon$. For our purpose here, both $\epsilon$ and $\Delta$ can be considered small so that the angles may be substituted for the tangents of the angle. Then, $\epsilon = h\Delta/(p - h)$. Substituting $\epsilon$ in the expression for $\Delta$ above, we obtain for the actual effective prismatic deviation

$$\Delta = \frac{p - h}{p} \Delta_o . \qquad \ldots (15)$$

As the object distance increases, the ratio $(p - h)/p$ approaches unity and the prismatic deviation for the eye becomes identical with the optical deviation $\Delta_o$.

*Example:* Suppose the object distance is 30 cm, and the distance of the prism from the center of rotation of the eye is 2.5 cm. Then $p - h = 27.5$ cm, and $(p - h)/p = 27.5/30 = 0.916$. Thus $\Delta = 0.916 \Delta_o$, which is equivalent to an 8.4 per cent decrease. A 10.00 prism diopter prism used for an object distance of 30 cm would actually produce a prismatic deviation at the eye of only 9.16 prism diopters. Obviously, the farther the prism is from the eye, the greater is this effect. The decrease in effective prismatic deviation becomes important in the phorometer, where the rotary prisms may be as much as 5.8 cm from the eye.

## 5. ABERRATIONS AND UNDESIRABLE PROPERTIES OF OPHTHALMIC PRISMS

Ophthalmic prisms for clinical use, either in measuring ocular deviations or in correcting oculomotor imbalances, have aberrations and other properties that are undesirable but in most cases unavoidable.

1. The variation of prismatic deviation with tilt of the prism has already been discussed.

2. Chromatic aberration due to the dispersive action of the prism, in which the blue rays are deviated more than the red rays, has also been mentioned. The result, however, is that the boundaries between dark and light areas, perpendicular to the base-apex meridian of the prism, appear to have color fringes not unlike small spectra. For low prismatic deviations this is unimportant, but patients may complain of these colored edges when prisms of large deviating power are used. There is no easy way to avoid these chromatic effects.

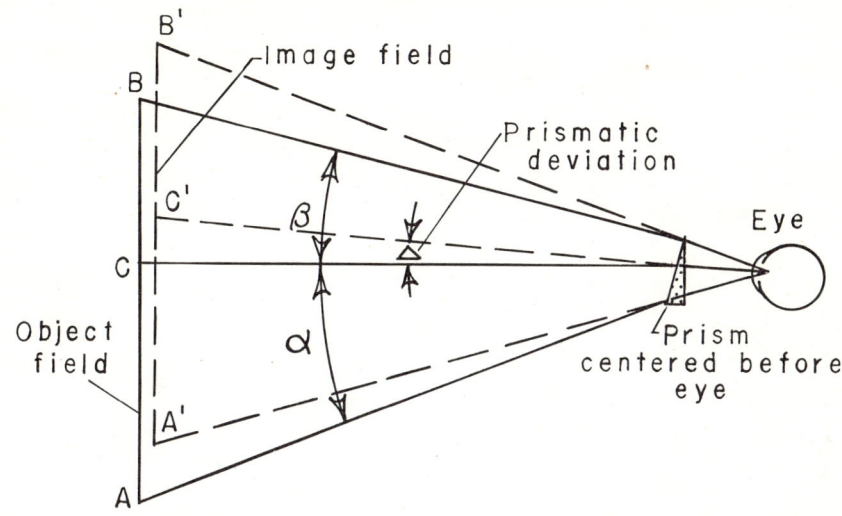

FIGURE 59. Maximal visual field seen through a prism.

3. Since the image of an object viewed through a prism appears displaced toward the apex, the field of view seen through the prism is decreased on the apex side of the prism. Figure 59 illustrates the fact that the visual field toward the apex of the prism is smaller than that toward the base. Because the entire image is displaced toward the apex of the prism, the extreme ray which passes through the apex, as compared with the extreme ray which passes through the base, must have originated closer to the center of the object field. Objects in the visual field that subtend angles on the apex side greater than the angle $\beta$ cannot be seen. The limiting factors are the size of the prism and the distance of the prism from the eye. In the case of prisms of the same size placed the same distance from the eye, the field toward the apex is progressively restricted as the prism power increases. This effect is especially noticeable when rotary or Risley prisms are used. To eliminate the effect, the entire prism should be decentered proportionately to the deviating power of the prism.

4. It is readily observable that the image of an extended object field seen through a prism is considerably distorted. This distorting effect is due to the fact that the cone or pencil

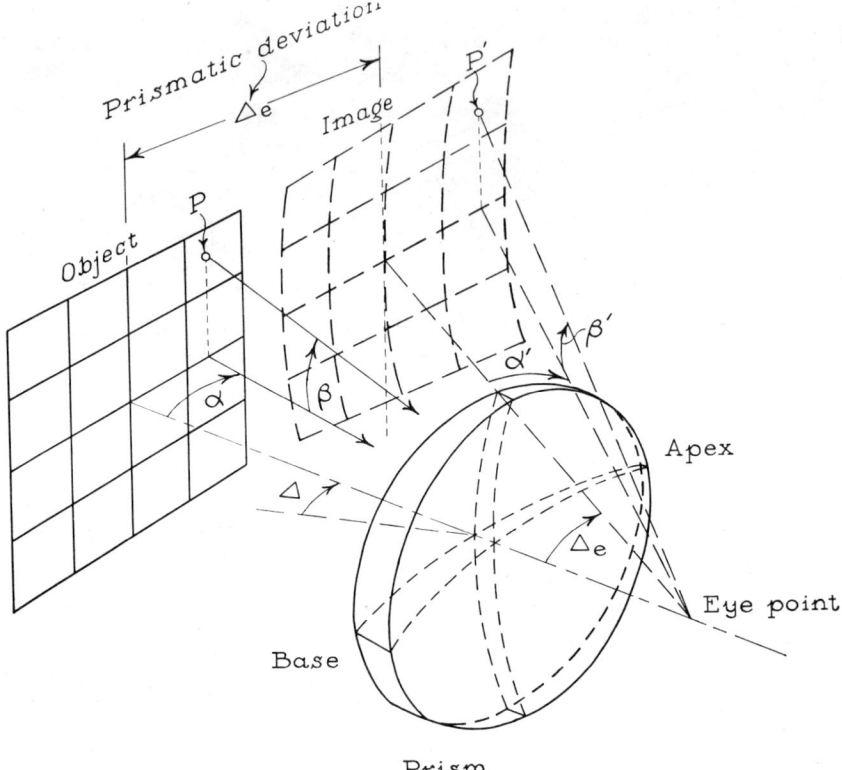

FIGURE 60. Distortion of the image of an extended object by an ophthalmic prism.

of light that passes through the pupil of the eye from each point in space traverses a different part of the prism, and the deviating effect of the prism upon this pencil of light will depend upon its particular angle of incidence and the length of its path through the prism.

There are three separate aspects of this distortion[1] (Fig. 60):

    a. The angular prismatic deviation increases toward the apex of the prism and decreases toward the base; that is, the image is increasingly expanded toward the apex and increasingly contracted toward the base of the prism. This is an asymmetric effect in the apex-base meridian of the prism—an asymmetric magnification of the image.

    b. There is a curvature of the images of all lines perpendicular to the apex-base meridian of the prism. This curvature also increases toward the apex of the prism.

c. There is a slanting of the image of all lines parallel to the base-apex meridian—those above and below the eye-level plane. The slant increases the farther a given horizontal line is from this plane.

The magnitude of the distortion of ophthalmic prisms can be changed by use of spherical (or cylindrical) surfaces for the faces of a given prism—by "bending" the prism. It can be shown that the asymmetric base-to-apex distortion just described in 4a can be virtually eliminated by the use of surfaces of a certain curvature. However, complete elimination of all distortion is not possible by this method.

This distortion of ophthalmic prisms, when prescribed for use in spectacles, often causes a false spatial localization in the stereoscopic perception of objects. The patient for whom prisms base-in are prescribed frequently objects to the apparent "bulging" of the print in the middle of the page.

## 6. RESOLUTION OF PRISMATIC DEVIATIONS

The question frequently is asked: If a prism is placed before an eye at an oblique axis, what are the prismatic deviations in the horizontal and vertical meridians? Since prismatic deviation is an angular displacement of the image in the apex-base meridian of the prism, that displacement can be treated as a vector—that is, graphically as a line whose length represents the magnitude and whose angle from the horizontal specifies the direction of the base-apex meridian. The prismatic deviation in any other meridian can be obtained by the usual method of resolving vectors.

Suppose a prism of deviating power $\Delta$ is placed before the right eye so that the base-apex meridian lies at an oblique angle, $\phi$—for example, base-in and base-down (Fig. 61). The question is, what are the prismatic deviations in the horizontal and in the vertical meridians? Practically, we can use a ruler and a protractor to find these, as shown in the figure. We first draw lines corresponding to the horizontal and vertical meridians and then, with the protractor, a line from the origin in the direction of the prism

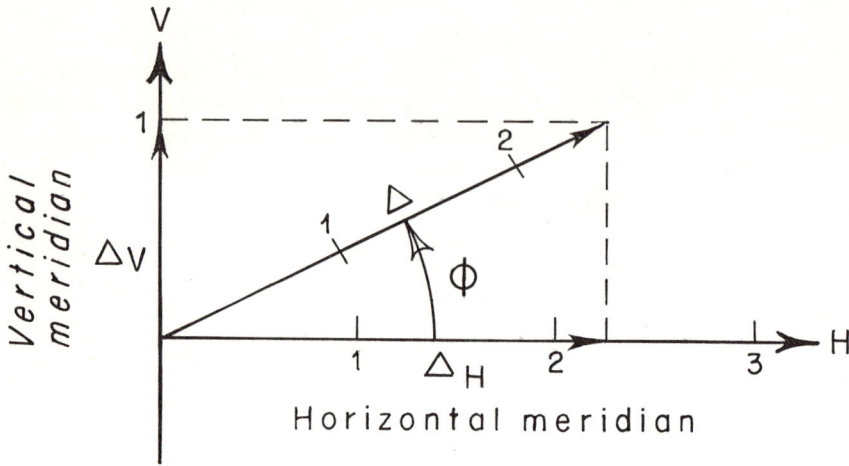

FIGURE 61. Type of graph used to find the vertical and horizontal component deviations of a prism whose meridian is placed at an oblique axis.

apex at the oblique angle $\phi$. Next we measure a length on this line corresponding to the power of the prism—say, by letting 2 cm correspond to 1 prism diopter. Now perpendiculars can be drawn from the end of this line segment to the horizontal and vertical meridians. The horizontal and vertical components, $\Delta_H$ and $\Delta_V$, can be found simply by measuring these projected distances with the ruler. Circular coordinate paper on which the concentric circles and radiating lines are printed would eliminate the need for the protractor and ruler.

The horizontal and vertical components of the prismatic deviation can also be computed by the simple trigonometric formula,

$$\Delta_H = \Delta \cos \phi \text{ and } \Delta_V = \Delta \sin \phi.$$

Examples of the magnitudes of the vertical and horizontal components for a prism of 1 prism diopter, when placed at different oblique meridians, are shown in the following table. When the prism is placed at the 45° meridian the horizontal and vertical components are equal. It will be clear from this table that the sum of the vertical and horizontal components of the deviation is *not* equal to the deviation in the base-apex meridian.

More frequently, however, we have the reverse problem: Namely, the horizontal prismatic deviation, $\Delta_H$, and the vertical

## Ophthalmic Prisms

| Angle φ | Vertical component $\Delta_V$ | Horizontal component $\Delta_H$ |
|---|---|---|
| 15° | .26 | .96 |
| 30° | .50 | .86 |
| 45° | .71 | .71 |
| 60° | .85 | .50 |
| 75° | .96 | .26 |

prismatic deviation, $\Delta_V$, are known; we need to learn how strong a prism placed at what meridian will give these deviations. The quantities can be found by computation from

$$\Delta = \sqrt{(\Delta_H)^2 + (\Delta_V)^2} \quad \text{and} \quad \tan \phi = \Delta_V / \Delta_H.$$

They also can be found graphically by laying out distances in the $H$ and $V$ meridians corresponding to the magnitude of the prismatic deviation in those directions. The resultant line, whose length corresponds to the desired prismatic deviation, is the hypotenuse of the right triangle formed by these lines. It can be measured with a ruler, and the angle of oblique meridian can be measured with a protractor.

An example is illustrated in Figure 62, in which the horizontal deviation $\Delta_H$ = 2.5 prism diopters base-in, and the vertical deviation $\Delta_V$ = 1 prism diopter base-up before the right eye. The equivalent single prism is one of 2.7 prism diopters at a meridian of 22°.

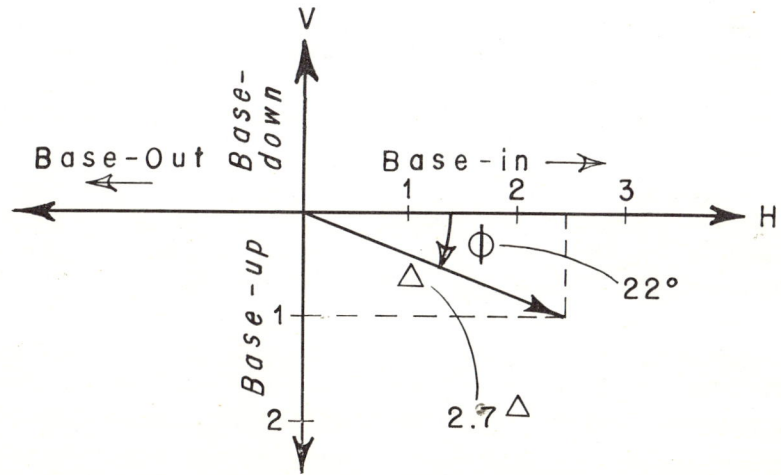

FIGURE 62. Graphic method of finding the resultant prismatic deviation and meridian angle for specified vertical and horizontal prismatic deviations.

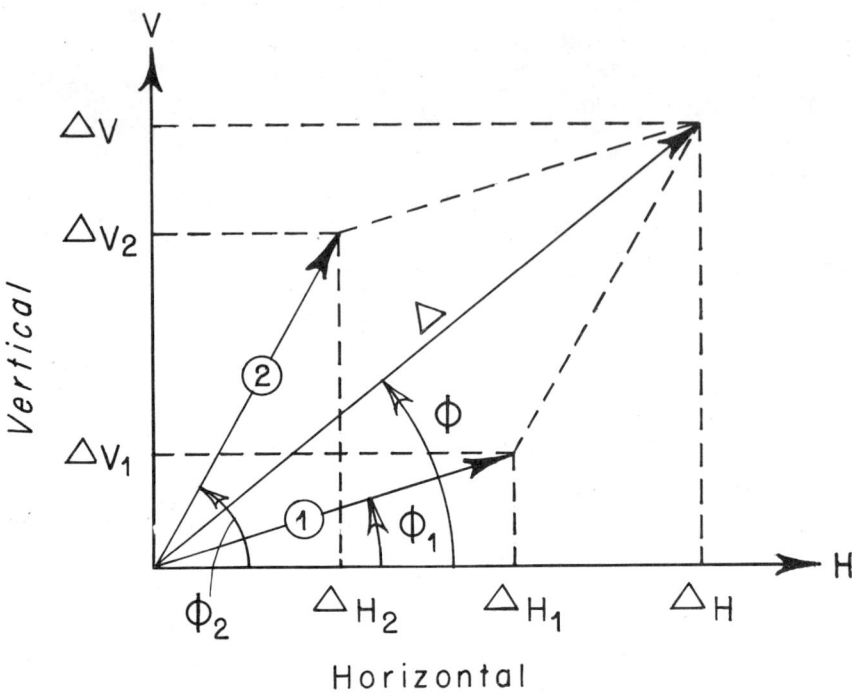

FIGURE 63. Graphic method of finding the resultant single prism which will give the same prismatic deviations as two separate prisms at different axes.

For even more complicated combinations of prisms, the same computational or graphic methods can be used. Suppose there are two prisms of unequal strengths placed at different oblique meridians. What single prism at what meridian gives the same horizontal and vertical components (Fig. 63)? The solution of the problem is the same as that used in high-school physics for finding the resultant of forces. The two lines representing each of the two prisms are drawn, using the scheme as shown in Figure 62. For each of these, the horizontal and vertical components, $\Delta_H$ and $\Delta_V$, are found. Then the sum of the two horizontal components is laid out on the horizontal meridian and likewise for the vertical meridian. From these the resultant and the meridian angle can be found easily.

The Risley or rotary prism is a good example of the way in which two prisms can be combined to give an optical unit for

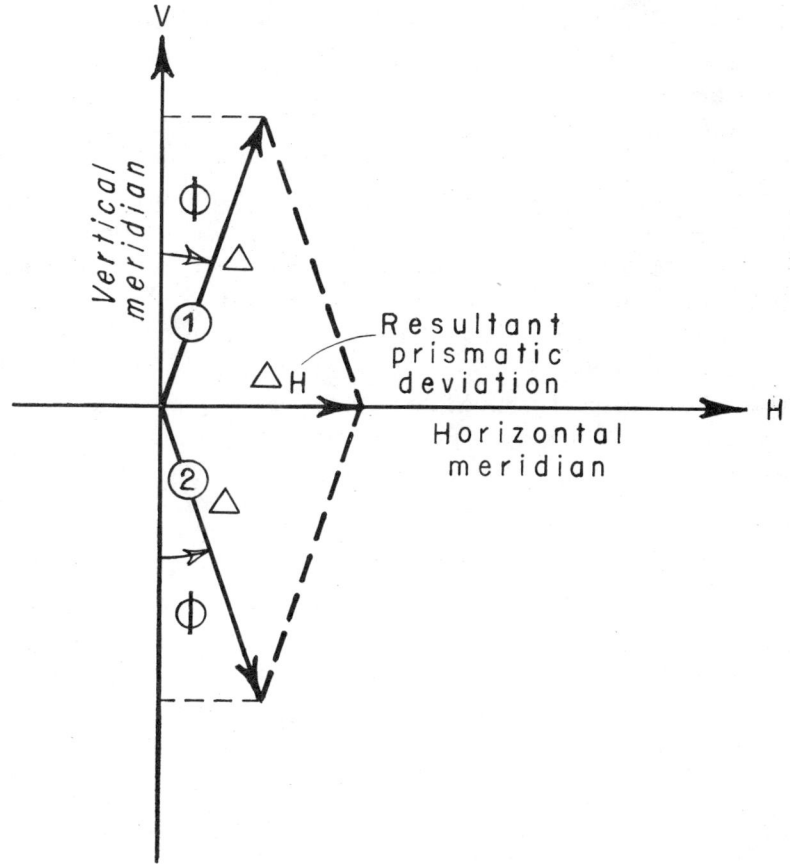

FIGURE 64. Graphic illustration of the resultant prismatic power of the Risley or rotary prism unit.

producing a continuously adjustable prismatic deviation in one meridian. Two matched prisms (matched for prismatic deviation $\Delta$) are placed together, the base of one opposed to the apex of the other, so that the prismatic deviation of the one completely neutralizes the prismatic deviation of the other. With the prisms in this arrangement, they function exactly as does a plane-parallel plate of glass. Let the two prisms be oriented so that the base-apex meridians are vertical. Now, by means of a circular rack-and-pinion device, the two prisms are rotated—one clockwise and the other counterclockwise—to exactly corresponding angles $\phi$. The resultant prismatic deviation can be represented graphically as shown in Figure 64. It will be seen that the vertical components of the two prisms are exactly equal but in opposite directions; hence there is no resultant vertical deviation. In the horizontal meridians the components of the two prismatic deviations are added, and the resultant prismatic deviation is entirely horizontal. In this particular case we can write that $\Delta_H = 2\Delta \sin \phi$. At the extreme rotation of 90°, the two prisms have bases in the same direction, and the maximal deviation of the combination of the two prisms is obtained—namely, $\Delta_H = 2\Delta$. If we continue to rotate the prisms still further in the same direction, the resultant prismatic deviation decreases. It should be noted that equal steps of rotation, $\phi$, do not result in equal steps of resultant prismatic deviation; hence the scale on the Risley prism is not linear.

## 7. THE BIPRISM

Nearly every trial set of optical lenses contains a biprism, which consists of two identical prisms placed base-to-base and made in one piece of glass (Fig. 65). A small source of light, $S$, is seen as two when the biprism is centered before the eye. This device is frequently used with a muscle light in the determination of heterophoria. When the biprism is placed before one eye—say, with the base-apex meridian vertical—the observer sees with that eye two vertically separated lights between which he sees with the other eye a single light. If a heterophoria is present, the three

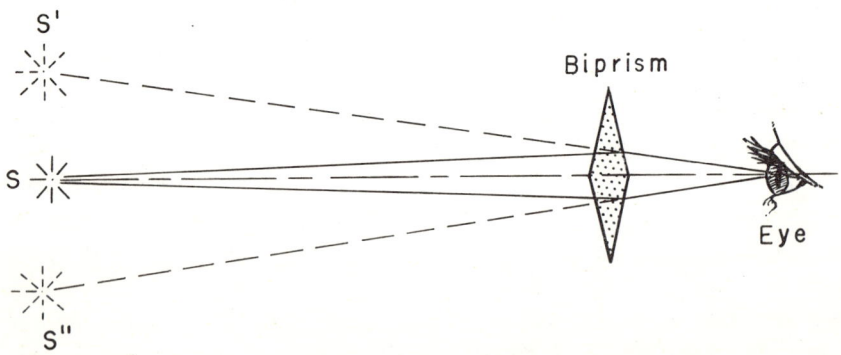

FIGURE 65. Biprism producing two images of a single test-light.

lights will not be aligned. The heterophoria can be measured by finding that prism, placed base-in or base-out, before the one eye, which will cause the three images to appear vertically aligned. The merit of the biprism is that both eyes receive the same stimulus to accommodation during the test. The disadvantage is the requirement that the prism be maintained fairly well centered with respect to the pupil of the eye.

## EXERCISES

1. What is the effective deviation produced by a rotary prism in the phorometer when the reading is 10 prism diopters for an object at 33 cm and the prism is 5.8 cm from the centers of rotation of the eyes?
2. If you were designing a Risley prism and wanted one with a maximal prismatic deviation of 10 prism diopters, what would be the powers of the matched component parts? What would be the prismatic deviation of this unit when these prisms had been rotated 30°? (The sine of 30° = 0.500.)
3. What are the power and meridian angle of a single prism that gives a prismatic deviation of 5 prism diopters base-in and 2 prism diopters base-down before the right eye?
4. A 5 prism diopter prism is placed with base-apex meridian at 30° from horizontal. What are the horizontal and vertical prismatic effects?

## REFERENCES

1. OGLE, K. N.: Distortion of the image by prisms. *J Opt Soc Amer, 41:*1023-1028, 1951.
2. PRENTICE, C. F.: *Ophthalmic Lenses, and Other Optical Papers.* Philadelphia Keystone Books, 1907, pp. 101ff.

*Chapter V*

# INTRODUCTION TO THE THEORY OF LENSES

## 1. INTRODUCTION—BASIC CONCEPTS AND DEFINITIONS

This section will be devoted to a discussion of a general theory of lenses and of the formation of images by lenses. The reader is encouraged to learn principles rather than to learn formulas—because if this suggestion is applied, whatever formulas are necessary will seem relatively simple and will follow readily from the principles involved.

First, it will be necessary to introduce certain special concepts pertaining to lenses and to define a number of terms that are constantly used in any discussion of the optics of lenses. These should be learned thoroughly.

### A. Vergence

The curvature of the wave front proceeding from a point source of light constantly decreases (becomes flatter) the farther the wave front is from the point source (Fig. 66). Near the point source the curvature is *great* because the radius of curvature is *small*. The farther a point, $P$, is from the source, $S$, the greater is the radius of curvature, and correspondingly the smaller is the curvature of the wave front at that point. At the point, $P$, the *vergence* of the light from $S$ is defined as the "curvature of the wave front," which in turn is defined as the *reciprocal* of the distance, $d$, from the source to the point.

Throughout the study of ophthalmic optics the distance $d$ from the source to a given reference point is expressed in *meters*, and the vergence of the light at the point $P$ is then said to be $1/d$ *diopters*. This unit proves to be a most convenient one in all oph-

# Introduction to the Theory of Lenses

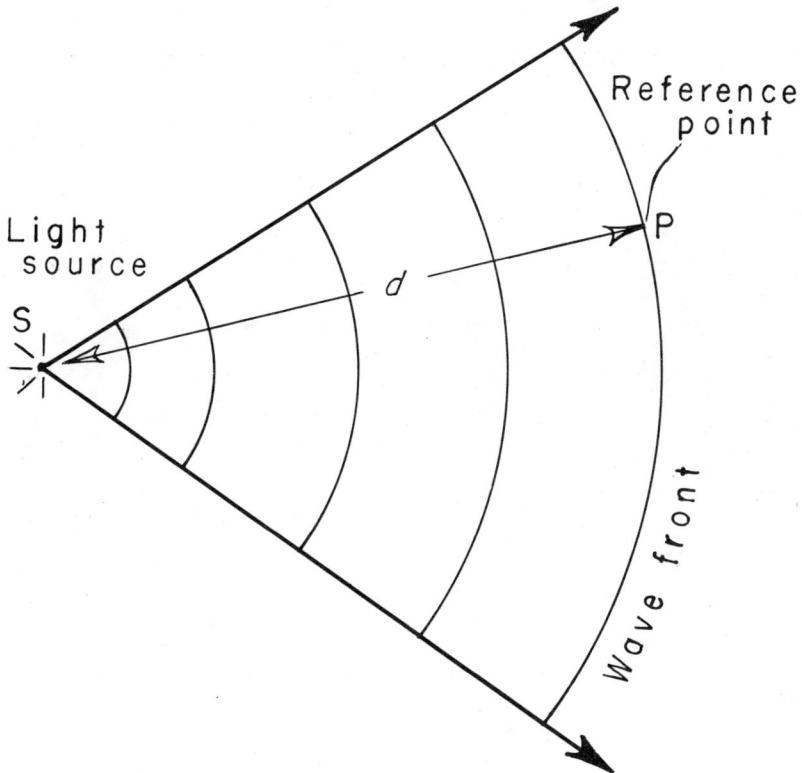

FIGURE 66. Diagram to introduce the concept of the vergence of light from a point source.

thalmic optical problems. Thus, in Figure 67, the vergence of the wave front of a pencil of light at a position, $B$, *1 meter* from the light source will be 1 diopter; at 2 meters (position $C$), 0.5 diopter; at 0.5 meter or 50 cm (position $A$), 2 diopters, and so forth. The farther the light source or object point is from the reference point, the more nearly the wave front approaches a plane. For an infinitely distant source the vergence $(1/d)$ will be zero. In that instance the rays perpendicular to the wave front will be parallel, and the light having such a plane wave front is often referred to as "parallel light."

Therefore, the vergence is represented (or measured) by the value $1/d$ diopters.

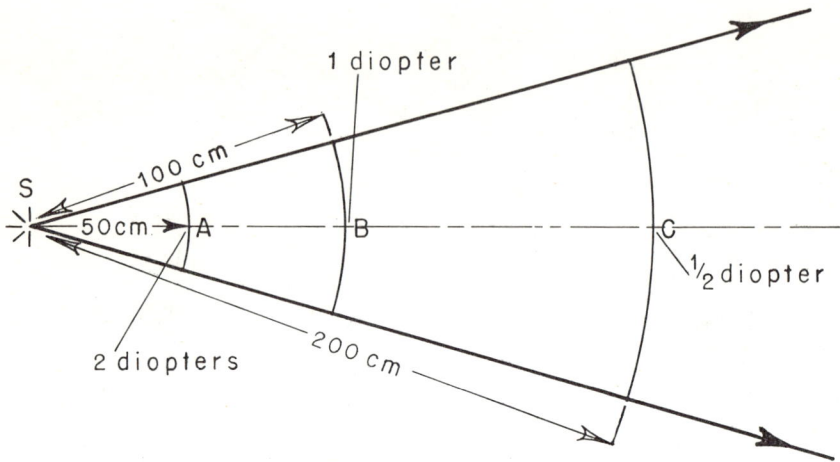

FIGURE 67. Diagram to define the unit of vergence—the diopter.

## B. Lenses

A *lens* is an optical device which changes by a constant amount the vergence of the light incident to and passing through it.* The effect of simple converging and diverging lenses on the vergence of the incident light may be visualized from the diagrams shown in Figure 68. Since light travels more slowly through glass than through air, the rays passing through the thick center of the converging lens are slowed down for a longer time than those passing near its thin edges. The rays of the emergent light, then, corresponding to the curved wave front, converge to a point $F$. Similarly, in the diverging lens the rays passing through the center are slowed down for a shorter time than those passing near its edges. The rays of the emergent light then diverge as if from a point $F$, to the left of the lens.

If the object point is nearer the lens, the rays incident on the lens will be diverging. Thus, in the case of the converging lens, the vergence of the emergent light will be less than that when the object point is more distant, and the rays will converge to the point $I$. In the case of the diverging lens, the lens will actually

---

*The exception to this definition is an afocal lens. There, however, the vergence changed by the first surface is counteracted by the change in vergence produced by the second (see page 147).

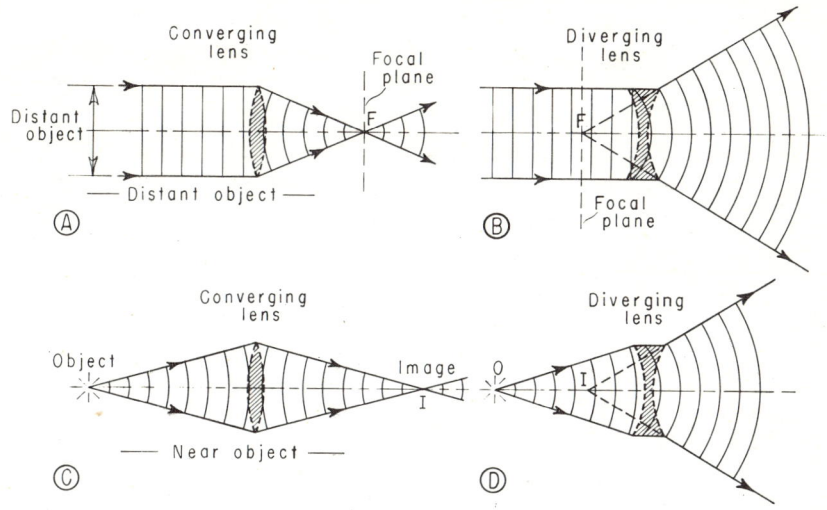

FIGURE 68. Changes in vergence of light as it passes through simple converging and diverging lenses.

increase the vergence of the emergent light, the rays then diverging as if from the point $I$.

If the direction of the incident light traversing the lenses is reversed—that is, from right to left—the wave fronts will be the same. This illustrates the principle of reversibility of light paths applied to lenses.

## C. Optic Axis

The optic axis (Fig. 69) of a simple lens is the straight line that passes through the centers of curvature of the front and back surfaces. A light ray that coincides with this axis passes through the lens undeviated.

## D. Focal Points

The focal point of a lens is that point on the optic axis at which all incident light rays parallel to the axis are brought to a focus—that is, a point to which they converge or from which they diverge. In the first two diagrams of Figure 68, in which the incident light is parallel, $F$ is the first or posterior focal point of the two lenses. There are, of course, two focal points for each lens, depending upon whether the parallel light is incident from the left or from

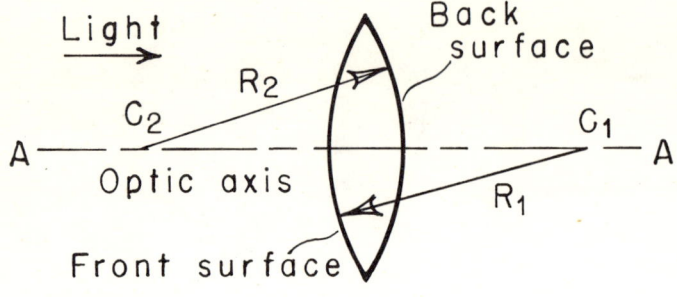

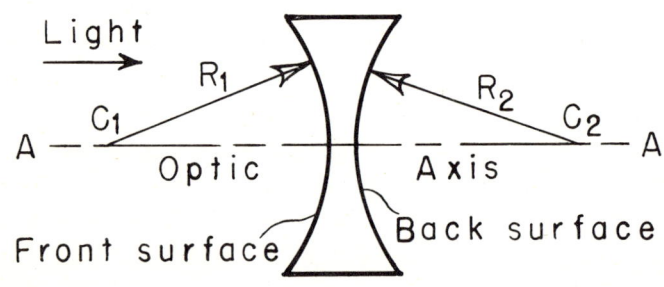

Figure 69. Diagram to illustrate the optic axis of a lens.

the right of the lens. If the light comes from the left, the focal point on the right side of the converging lens (or on the left side of the diverging lens) is called the *first* focal point. If the light is incident from the right, the focal point is called the *second* focal point. From the opposite point of view, the first focal point corresponds to the position (on the axis, in front of the lens) of a point source whose rays, when emerging from the lens, are parallel. For the eye, these focal points are called the *posterior* (near retina) and *anterior* (in front of the cornea) focal points, respectively.

## E. Focal Plane

The focal plane is a plane erected perpendicularly to the optic axis at the focal point. In an ideal lens, parallel rays incident to the lens at an oblique angle come to a focus at some image point on the focal plane (Fig. 70).

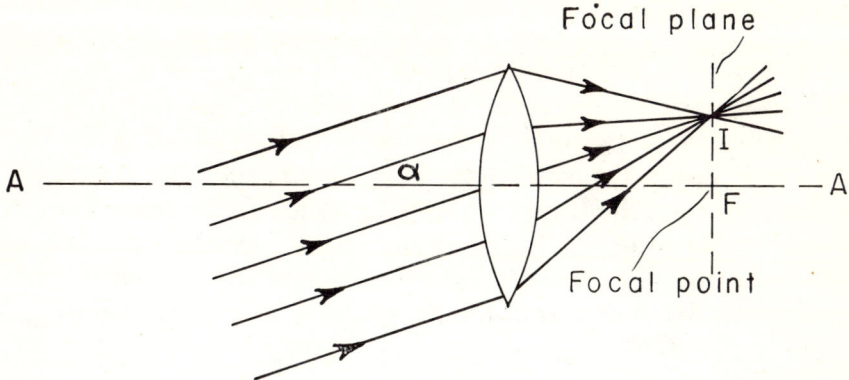

FIGURE 70. The focal plane, at the focal point and normal to the optic axis.

### F. Paraxial Rays

Paraxial rays are those incident on the lens which lie close to the optic axis. This term is used somewhat loosely. For the lens designer, rays incident on the lens near the axis have angles of incidence and of refraction that are small, so that the angles themselves can be used in calculations instead of their sines (as in Snell's law).

### G. Extra-axial Rays

All rays incident to the lens which are not paraxial are said to be extra-axial. Usually, such rays are assumed to lie in a plane that also includes the optic axis. A *skew ray* is one incident to the lens which does not lie in a plane that includes the optic axis and therefore does not cross the optic axis either before or after passing through the lens.

### H. Chief Rays

In a pencil of light, or bundle of rays from an object point, incident upon a refracting surface, a central ray of the pencil can be selected as representative of the subsequent course of the pencil. This ray is called the *chief ray*. Usually, the chief ray is the one directed to the center of the pupil, or the aperture, of the optical system.

## I. Power

The power of a lens is an expression of the degree to which it can change the vergence of the incident light. This quantity (in diopters) is roughly equal to the reciprocal of the distance (in meters) of the focal point from the lens. This is logical, for (referring to Fig. 68) when the vergence of the incident light is zero (its rays being parallel), the resulting vergence of the emerging light is due entirely to the lens, and the rays converge to (diverge from) the focal point. More details of the definition will be given later.

## 2. THE IMAGE FORMATION BY AN IDEAL INFINITELY THIN LENS

The principles of image formation by all types of lenses can be best understood by a consideration of an ideal aberrationless lens assumed to be infinitely thin.[1] From this the problems concerned with the more complex thick lenses and lens systems can be approached more easily.

### A. Graphic Method

For this ideal lens we can diagram the course of particular rays from an object point through the lens and can thus locate the position of the image formed by the lens. This method of diagramming should be fully understood.

Consider, first, an ideal converging lens (Fig. 71) represented by the line $L—L$. The line $A—A$ will be the optic axis perpendicular to the line representing the lens. The points $F_1$ and $F_2$ will be the first and second principal focal points. Light rays from an infinitely distant light source on the axis, which are therefore parallel to the axis, incident on the lens from the *left*, will all converge to the point $F_1$. Similarly, if the light rays were to come from an infinitely distant light source and were incident on the lens from the *right*, they all would converge to the point $F_2$ on the axis. This lens is in air, and the distances of these two focal points from the lens are equal; that is, the first and second focal distances are equal.

Again we shall plan the diagram so that the direction of the light will be from left to right, and we shall place the real object to the left of the lens.

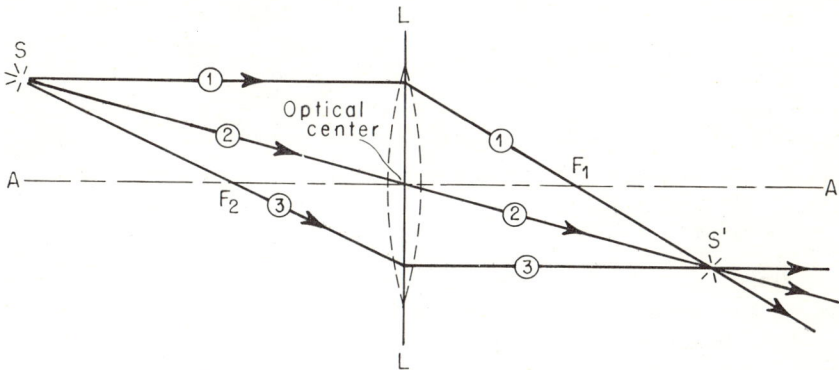

FIGURE 71. An ideal thin lens and the three selected rays from an off-axis object point used to locate the position of the image point.

Suppose now that in the same diagram (Fig. 71) the point $S$, to the left of the lens, is an off-axis object point or a point source of light. To locate the position of the image of $S$ formed by the lens, we can use three lines representing particular rays from among all those arising from $S$, though only two of them are really necessary. *First*, draw a line (1) from the object $S$ parallel to the optic axis to the plane of the lens. This line represents a ray parallel to the axis as if it were coming from a very distant object, and on passing through the lens it will be deviated so as to pass through the principal focal point, $F_1$, of the lens. *Second*, draw a line (2) representing that ray from $S$ directed to the center of the lens. Just as a ray coinciding with the axis is undeviated, so also all rays—including this one—directed to the optical center of the ideal lens are undeviated. The intersection of these two lines (1 and 2) determines the position of the image, $S'$. A third line (3) can also be drawn representing that ray from the object point, $S$, directed to the second or anterior focal point, $F_2$, but continued to the point at which it intersects the lens. The ray represented by this line must then emerge from the lens parallel to the axis, and the line so drawn will also intersect the first and second lines at the same image point, $S'$. Clearly, only two of these lines are necessary for diagramming the geometric position of the image, but the third is a useful check.

If the object point $S$ lies on the axis of the lens itself, the cor-

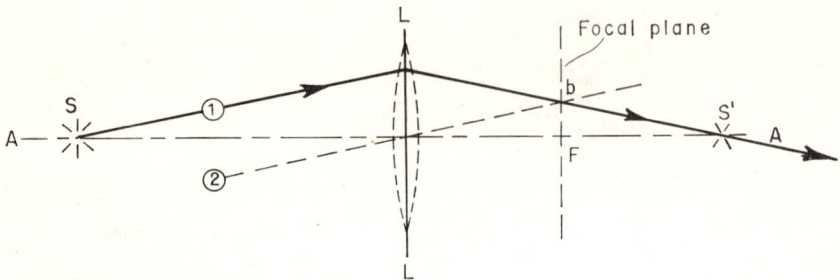

FIGURE 72. Method of diagramming rays to locate the position of the image of an object on the optic axis of an ideal converging lens.

responding image can be found by consideration of the diagram in Figure 72. First, *any* line (1) is drawn from the object point, $S$, to the lens plane. Next, a line (2) parallel to line 1 is drawn which passes through the center of the lens. This line, representing a ray, is undeviated. Now, since the lens represented is an ideal one, all parallel rays incident on the lens must emerge converging to an image point on the *focal plane* of the lens (Fig. 70). The intersection of line 2 with the focal plane determines a point, $b$, which is the image point for all incident rays parallel to line 2, including line 1. Therefore line 1, as an emergent ray, is drawn through $b$ until it intersects the axis. This point of intersection locates the image, $S'$, of the axial object point, $S$, since the axis corresponds also to an undeviated ray from $S$ to $S'$.

Figure 73 illustrates the method for diagramming the rays in the case of a *diverging* lens to find the position of the image for a given object point. Parallel rays incident to such a lens diverge after passing through it *as if originating* from some point on the first *focal plane*. The same procedure of diagramming as that used for the converging lens is followed. From the object point, $S$, a line (1) is drawn parallel to the axis of the lens. Like the corresponding ray, on emerging it deviates outward as if it had originated from the first focal point, $F_1$.

Second, a line (2) is drawn from $S$ through the center of the lens. Like its ray, this line is undeviated on passing through the lens. The intersection of these two lines projected backward determines the position of the image, $S'$, which is on the same side of the lens as is the object, $S$. A third line (3) also can be

*Introduction to the Theory of Lenses* 91

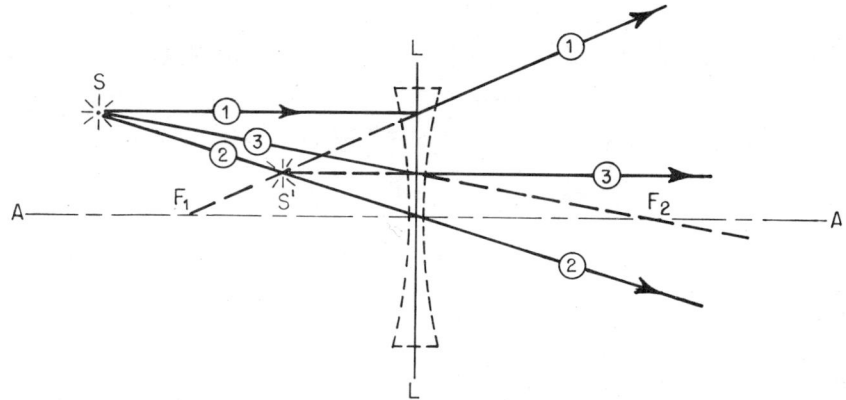

FIGURE 73. Method of diagramming rays to locate the position where a thin ideal diverging lens forms the image of an off-axis object point.

drawn from the object point, $S$, directed to the second focal point. A ray so directed would emerge from the lens in a direction parallel to the axis of the lens; this line extended will also intersect the other two lines at the image, $S'$.

In the preceding diagrams (Figs. 71, 72, and 73) the points $S$ and $S'$ are called *conjugate* points.

If either is the object, the other will be the image, by the principle of reversibility of the direction of rays. So that one may be familiar with the procedure outlined above, typical situations for both converging and diverging lenses should be diagrammed. For examples, consider the cases in which the object point lies (A) outside the focal point, (B) at twice the focal length, (C) inside the focal length, and (D) at the second focal point. Figure 74 schematically shows diagrams for situations A, B, and C.

## B. Analytic Method

The quantitative relationships between the distances of the object and the image measured from the thin lens now can be derived in terms of the focal length of the lens. First, for a converging lens, let us redraw the diagram in Figure 71 and specify the dimensions as shown in Figure 75. As before, the object distance from the lens is taken as $u$, the image distance as $v$, and the focal length—the distance of the first focal point from the lens—

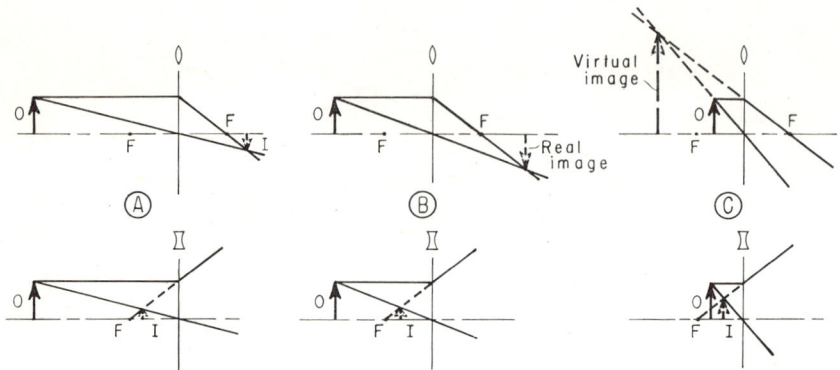

FIGURE 74. Positions of the image produced by a converging and a diverging ideal thin lens for different positions of the object.

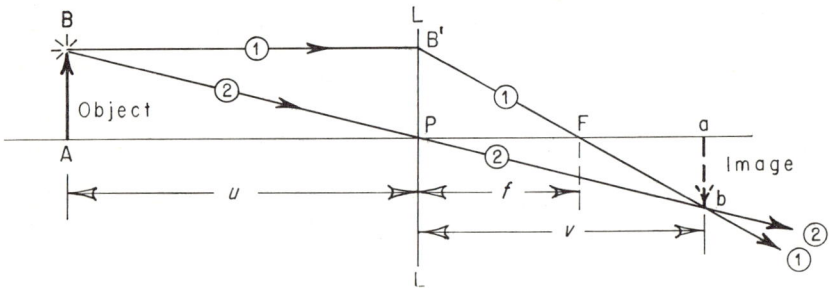

FIGURE 75. Diagram used in deriving the relationship between object and image distances and the focal length of a thin lens, also illustrating the concept of magnification of the image.

as $f$. The figure has been drawn so that all of these will be plus quantities.

To derive the formula giving the relationship between the image and object distances, we make use of similar triangles formed by the various lines representing particular light rays. Consider first the two right triangles, $APB$ and $aPb$, pertaining to the object and image. These two right triangles are similar because the included angles are equal. We can write, then,

$$\frac{ab}{AB} = \frac{I}{O} = \frac{v}{u}, \qquad \ldots (16)$$

in which $I$ and $O$ are the sizes of the image and object, respectively. Consider, now, the two similar right triangles, $aFb$ and $PFB'$, for which the included angles also are equal, and keep in mind that $PB' = AB = O$, the size of the object. Thus

$$\frac{I}{O} = \frac{v - f}{f}.$$

Equating these two expressions for $I/O$, we have

$$\frac{v}{u} = \frac{v - f}{f},$$

which can be put in a standard form,

$$\frac{1}{v} + \frac{1}{u} = \frac{1}{f}. \qquad \ldots (17)$$

From equation 16 the magnification $M = I/O$ is given by

$$M = -\frac{v}{u}. \qquad \ldots (18)$$

In the above expression for the magnification the minus sign is arbitrarily inserted to indicate that the image of the object is inverted.

These equations (17 and 18) are the fundamental relationships that hold for all perfectly ideal, infinitely thin lenses and for paraxial rays for all real lenses. In many optical problems these relationships will hold if the lens can be considered thin.

The equations also hold for diverging lenses, except that $f$ must be taken minus because the first focal point is to the left of the lens. Such lenses are more often referred to as negative or minus lenses.

The dioptric *power* of these thin lenses is defined as the "reciprocal of the focal length," namely, $F = 1/f$.

Now, equation 17 essentially states that the *vergence* of the emergent light, compared to that of the incident light, is equal to the converging (or diverging) power of the lens. This follows because $1/u$ is the vergence of the incident light, and $1/v$ is the vergence of the emergent light. If we write $U = 1/u$ and $V = 1/v$, which states *vergences* of the incident and emergent pencils of light directly, then equation 17 can be written simply

$$V + U = F,$$

and accordingly

$$M = \frac{I}{O} = -\frac{U}{V}. \qquad \ldots (19)$$

If all distances, $u$, $v$, and $f$, are measured in meters, the vergences and the power of the lens will be in diopters by definition. This procedure then reduces our optical imagery problem to simple processes of addition and subtraction of vergences.

*Note:* Strictly speaking, the vergence of the light from an object to the left of a lens will be diverging and therefore minus. In order to avoid confusion and to be consistent with most optical problems met in ophthalmic optics, we are in essence arbitrarily making this vergence minus in our derivations.

*Example 1:* A converging lens has a focal length of 25 cm. An object, the size of which is 5 cm, is placed 100 cm in front of the lens. What is the image distance? What is the magnification and what is the size of the image?

Given are $u = 1.00$ meter; $f = +0.25$ meter; and the object size, $O = 5$ cm. Then, $U = +1.00$ diopter, and $F = +4.00$ diopters. It follows that $V + 1.00 = +4.00$, or $V$ (the vergence of the emergent pencil of light) $= +3.00$ diopters; thus $v$ (the distance of the image from the lens) $= 0.333$ meter or 33.3 cm. The magnification $M = -33.3/100 = -0.333$. Thus, the size of the image $I = MO = -(0.333)(5) = -1.66$ cm.

*Example 2:* A diverging lens has a focal length of 25 cm. The size of an object 100 cm from the lens is 5 cm. What is the image distance from the lens? What is the magnification? What is the size of the image?

Given are $u = 100$ cm; $f = -25$ cm; and the object size, $O = 5$ cm. We have, then, $F = -4.00$ diopters and $U = +1.00$ diopter. Accordingly, $V + 1.00 = -4.00$, whence the vergence of the emergent light $V = -5.00$ diopters, whence the image distance is $-1/5.00 = -0.200$ meter $= -20$ cm. The minus sign found for $V$ and $v$ shows that the image is in front of the lens, on the same side (the left) as that of the object, and the image is virtual. The magnification $M = -I/O = -(-.20)/1.00 = +0.20$, when the size of the image $I = +1.00$ cm. The fact that $I$ has a plus sign indicates that the image is erect.

*Example 3:* Suppose one observes an object 33.3 cm from a lens in front of the eye and wishes to see a virtual image of that object at a distance of 5 meters from the lens. What should be the power of that lens, and what will be the magnification of image to object?

Given are $u = +33.3$ cm and $v = -5$ meters. Then, $U = +3.00$ diopters and $V = -0.20$ diopter. Thus, $-0.20 + 3.00 = F$, or the power of the lens $F = +2.80$ diopters. The magnification $M$ would be $5/0.333 = 15$.

We cannot overemphasize the importance of thorough familiarity with the method of diagramming the relationship between the object and the image of a simple lens. The foregoing three examples should be diagrammed. It is far less important to remember the equations than it is to remember how to draw these descriptive rays and to visualize the change in vergences. The equations can always be derived from the diagram.

## EXERCISES

1. What are the vergences for light at points at the following distances from a light source: 100 cm, 33.3 cm, 400 mm, 25 cm, 0.50 meter, 100 mm?
2. It is found that the image of a distant scene, limited by the light passing through a window, can be focused sharply on a white card held 40 cm from a lens. What is the power of the lens?
3. If one holds a lens 50 cm from a flash lamp and finds that an image of the lamp is sharply focused when the card is 16.6 cm from the lens, what is the power of the lens?
4. A circular object 4 cm in diameter and 40 cm from a lens has an image 2 cm in diameter. What is the distance of the image from the lens, and what is the power of the lens?
5. An object 5 cm in diameter is 33.3 cm from a converging lens of 8 diopters. Calculate the position of the image from the lens and its size.
6. An object 5 cm in diameter is 33.3 cm from a diverging lens of 8 diopters. Calculate the position of the image from the lens and its size. Is this image real or virtual? Why?
7. An object 4 cm in diameter is placed 167 mm from a converging lens of 5 diopters. Calculate the position of the image from the lens and its size. Is this image real or virtual?

## REFERENCE

1. SOUTHALL, J. P. C.: *Mirrors, Prisms and Lenses.* New York, Dover, 1964, pp 226ff.

Chapter VI

# SOME APPLICATIONS OF THIN-LENS THEORY TO VISUAL OPTICS

$M$ANY aspects of visual optics and ophthalmic optics can be readily understood by applying the elementary principles learned from the study of the formation of images by thin lenses. This follows because in many cases we can consider the lens systems as thin lenses, with reasonable accuracy. In certain problems even the refractive system of the eye itself can be considered as if it were collapsed to a thin lens. Hence, it is convenient to take up now a number of the problems in visual optics, making use of this thin-lens theory, before actually discussing the formation of images by spherical and cylindric surfaces or the principles of thick lenses.

## 1. ANGULAR MAGNIFICATION

The magnification of the image relative to the object for a given optical system is defined as

$$M = \frac{\text{size of image}}{\text{size of object}} = \frac{I}{O},$$

and this ratio has been shown to be equal to the ratio of the image and the object distances, $v$ and $u$, respectively. This statement is the strict definition of magnification. Even when the size of the image actually is smaller than the size of the object, in which case $M$ is less than one, the ratio is still called the magnification.

When an optical instrument is used in conjunction with the eye, however, we must deal with a different aspect of magnification, namely the comparison of the *angular* size of the image of an object as seen through the instrument to the angular size of that object as seen without the instrument (Fig. 76). We are concerned, then, with the *angular magnification*, which must refer to some *eye point*, such as the entrance pupil of the eye or, in the case of eye movements, to the center of rotation of the eye itself. (For the present,

Some Applications of Thin-Lens Theory to Visual Optics 97

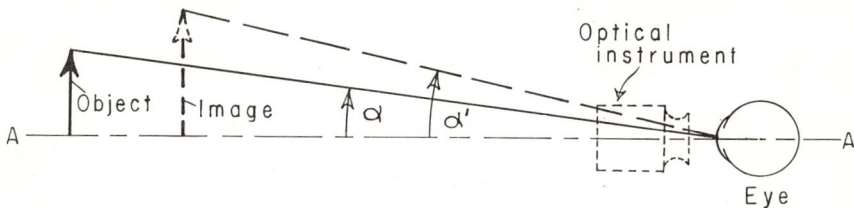

FIGURE 76. Diagram illustrating the definition of angular magnification produced by an optical system.

the eye is assumed to be without refractive error.) The angular magnification $A$ is defined as

$$A = \frac{\text{angle subtended by the image seen through instrument}}{\text{angle subtended by the object seen without instrument}},$$

referred to some specific eye point.

## A. The Magnifying Lens

Consider first the ordinary hand magnifier as a simple, infinitely thin lens—which it is not, of course, but for the purpose here it is used as if it were. Ordinarily, such a lens is held so that the image of an object seen through it is erect and magnified. This means that the object must lie just inside the second focal plane of the converging lens. Figure 77 shows schematically the relationships between the object, lens, and image. To avoid the problem of signs, we can redraw the figure so that all quantities are plus and $u$ is greater than the anterior focal distance (Fig. 78). Let the power of the lens be $F$ diopters, the distance of the lens from a reference point in the eye be $h$ (meters), and the distances of the object and the image from the lens be $u$ and $v$ (meters), respectively. The angular size of the object without the magnifier will be $\alpha$,

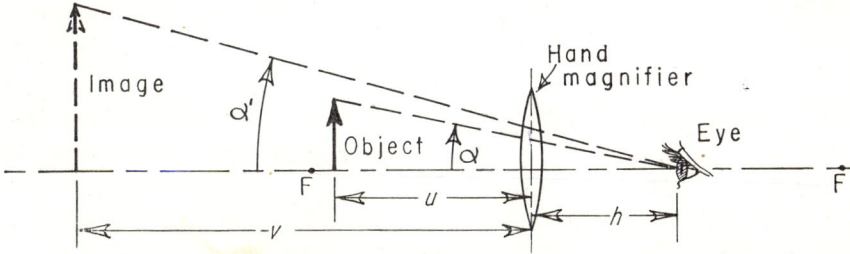

FIGURE 77. Optical principles of an ordinary hand magnifying lens.

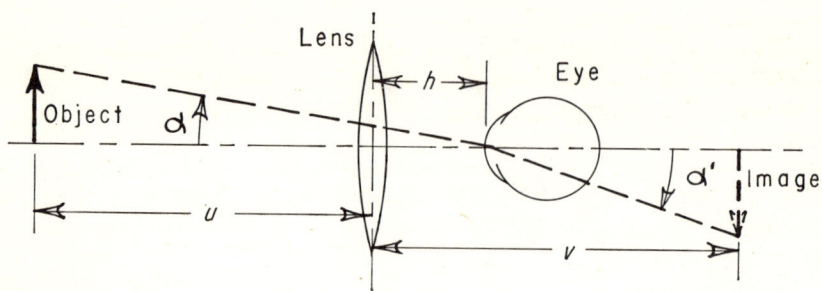

FIGURE 78. Diagram used in deriving the formula for angular magnification.

and the angular size of the image of the object seen through the magnifier will be $\alpha'$. Clearly, $\tan \alpha = O/(u + h)$, and $\tan \alpha' = I/(v - h)$. Remembering that $I/O = v/u$, we can write for the angular magnification

$$A = \frac{\tan \alpha'}{\tan \alpha} = \left[\frac{I}{O}\right]\left[\frac{u + h}{v - h}\right] = \left[\frac{v}{u}\right]\left[\frac{u + h}{v - h}\right] = \frac{1 + h/u}{1 - h/v}.$$

If vergences are used, $1/u = U$ and $1/v = V$; then remembering that $V + U = F$, we obtain on substitution for the angular magnification

$$A = \frac{1 + hU}{1 - h[F - U]}. \qquad \ldots (20)$$

In this situation, $U$ is always slightly greater than $F = (1/f)$; that is, $u < f$, and the object lies just inside the second focal point of the lens. The denominator in this ratio will be nearly one because $[F - U]$ differs little from zero. Optimal magnification occurs when $U = F$—that is, when the object is placed at the anterior focal point of the lens, whereupon the image distance becomes infinite (the accommodation of the eye looking through the lens must be completely relaxed to see the image clearly). Then the maximal magnification, $A_o$, will be

$$A_o = 1 + hF. \qquad \ldots (21)$$

This formula for the angular magnification refers to the ratio of the angular size of the image of the object seen through the lens to the angular size of the object seen without the lens, the object remaining at the *same distance* from the eye in both circumstances.

*Example:* Suppose a person views reading material at a distance of 25 cm from the eye and then holds an ordinary hand magnifying lens with a focal length of 12.5 cm (and therefore a power of 8 diopters) above the print at the farthest distance at which a clear image can be obtained with relaxed accommodation (then $u = f$). What is the angular magnification? Given: The power of the lens, $F = 8$, and $u = 12.5$ cm. The distance of the lens from the eye is $h = 25.0 - 12.5 = 12.5$ cm $= 0.125$ meter. Thus, the angular magnification is $A_o = 1 + (0.125)(8) = 1 + 1.0 = 2.0$. The angular size of the print, therefore, is enlarged two times (2X).

## B. Subnormal Vision Lenses

This discussion also is applicable to the use of convex lenses of relatively high power placed close to the eye to aid reading by persons with subnormal visual acuity.[1]

Suppose a +10.00-diopter converging lens is placed 2 cm from the cornea of the eye. A page of print is then brought toward the eye up to the second focal point of that lens so that the eye sees through it a virtual image of the print at an infinite distance. The angular magnification produced by the lens would be $A_o = 1 + hF = 1 + (.02)(10) = 1.20$, which is only 20 per cent [per cent magnification $= 100 (M - 1)$]. This magnification of course refers to the print held before the eye at a distance of $10 + 2 = 12$ cm, with and without the lens. Without the lens the image of the print would be very blurred.

Usually, the angular magnification for such a lens as specified refers to the reading card being held at a normal reading distance of, say, 40 cm. When viewed through the lens the card is brought to a position close to the eye—that is, to the second focal point of the lens. Consequently, the greater part of the angular magnification obtained is due, not to the effect of the lens, but to the fact that the card is brought nearer the eye (Fig. 79). The angular

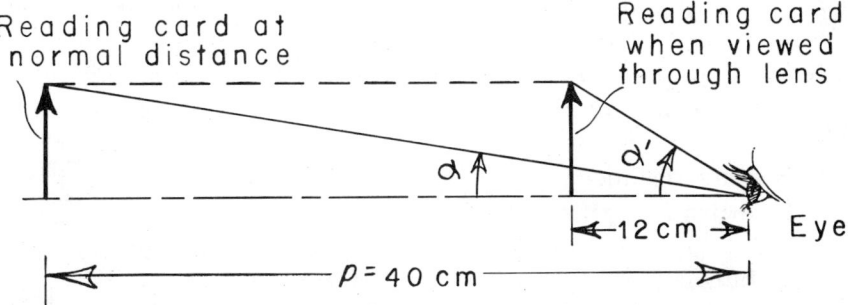

FIGURE 79. Moving a test card nearer the eye increases the angular magnification.

magnification is the ratio of the two distances. The distance of the card from the eye when viewed through the lens (in the example) would be 12 cm. Therefore, this part of the angular magnification is 40/12 = 3.33, or more than three times. The total angular magnification obtained with the use of the lens but referred to the distance of 40 cm is the product of the two separate magnifications, or $A$ = (3.33) (1.20) = 4.0, or four times. If the visual acuity of the eye without the lens was effectively 20/200, then with the lens the acuity would be effectively 20/50 for the print.

If $p$ is the reference viewing distance, and $(f + h)$ is the distance of the object when viewed through the lens, then the angular magnification will be

$$A = \left[\frac{p}{f+h}\right]\left[1 + hF\right] = \left[\frac{pF}{1+hF}\right]\left[1 + hF\right],$$

or

$$A = pF. \qquad \ldots (22)$$

The angular magnification, then, of a so-called microscopic ophthalmic lens for subnormal vision is equal to the power of the lens multiplied by the *reference* distance in meters. In the example here $A$ = (0.40) (10) = 4.0. This relationship clearly points up the importance of stating the reference viewing distance—though this is somewhat arbitrary, since the object is never actually viewed at this distance. As an historical carry-over from the angular magnification of microscopes, this reference viewing distance or working distance ("minimal distance of distinct vision") is frequently taken as 25 cm. How these lenses for subnormal vision should be specified, then, depends on an agreement in choice of this reference distance. For ophthalmic optics the normal reading distance of 40 cm would be preferable to the 25 cm so often used.

It must be pointed out that the two relationships for angular magnification in equations 21 and 22 give the same result when $h = p - f$. This is the case when the object (print) remains at the reference distance $p$, and the lens is used as a hand magnifier held at a distance $f$ from the object. The practical disadvantage of using the lens in this way to obtain the higher magnifications needed by subjects with subnormal vision is that the size of the visual field is limited by the diameter of the lens. Although larger lenses could be used, they would be heavy, and (even with aspheric surfaces) the aberrations would still tend to restrict the useful visual field.

### C. Magnification of the Ophthalmoscope

The angular magnification of the ophthalmoscope is another example to which the above discussion applies. The dioptric system of the subject's eye can be considered as a single converging lens, through which the examiner views the retina. Since the retina is very close to the first focal point of this "lens," we can think of the lens as a hand magnifier, which would place the image of the retina at a great distance behind the eye being observed (see Fig. 80). The examiner uses relaxed accommodation to view this image.

In specifying the angular magnification through the ophthalmoscope, again some reference distance must be selected. Since the power of the optical system of the eye is about 58 diopters, if we take the usual 25 cm, the angular magnification is $A = (0.25)(58) = 14.5$. This means merely that the angular size of the structures of the retina being examined through the ophthalmoscope is 14.5 times larger than the angle those same structures would subtend at the examiner's eye if the subject's retina itself (with optical structures removed) were observed at a distance of

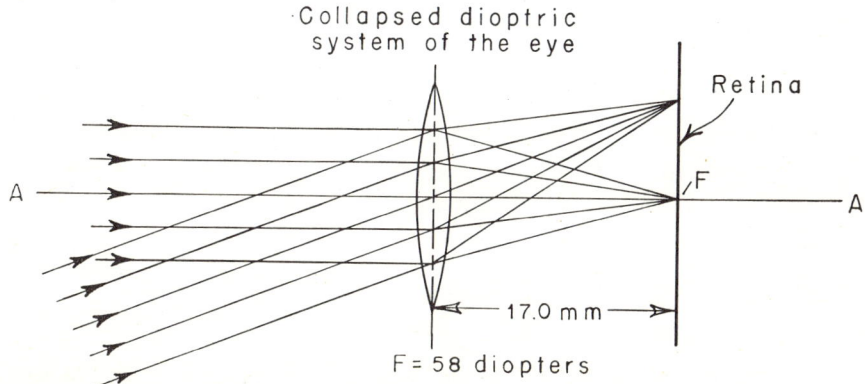

FIGURE 80. The dioptric system of the eye acts as a hand magnifier so that the retinal structure can be observed (see also Fig. 156).

25 cm from the examiner's eye. Obviously, this is a relative magnification, for the choice of the reference distance, 25 cm, was arbitrary.

The more general formula for the angular magnification of a magnifier referred to some arbitrary reference distance, $p$, when the object does not coincide with the focal point of the lens, is

$$A = pU/[1 - h(F - U)]. \qquad \ldots )23$$

All these formulas will be slightly changed when thick lenses are used because then another small magnification factor due to the thickness and the shape of the lens must be included.

This discussion of angular magnification does not apply directly to the magnification of the retinal image when the optical instrument before the eye is an ophthalmic lens that corrects a refractive error. That problem is more complex and will be considered later in this monograph.

## 2. PRINCIPLES OF OCULAR REFRACTION

Some general propositions that apply to ocular refraction can be considered here, again using the simple thin-lens theory to replace the more complicated optical system of the eye (which will be dealt with later).

In the following discussion, each optical system will be assumed ideal—that is, consisting of thin lenses. This assumption is indeed an oversimplification, especially when applied to the dioptric system of the eye.

### A. Emmetropia

If the accommodation of an eye is relaxed and the parallel rays from a very distant object are focused on the retina, the eye is said to be *emmetropic;* then visual acuity is at maximum. The *far point*, namely that object point on the visual axis whose image is conjugate to the retina and therefore is sharply focused on it, is at infinity.

If the dioptric power of the eye remains unchanged when the object is brought closer, the vergence of the incident light reduces the effective power of the eye so that the image of the object lies beyond the retina (Fig. 81). The rays of light, in being intercepted by the retina, would form a blur instead of a sharp image.

# Some Applications of Thin-Lens Theory to Visual Optics 103

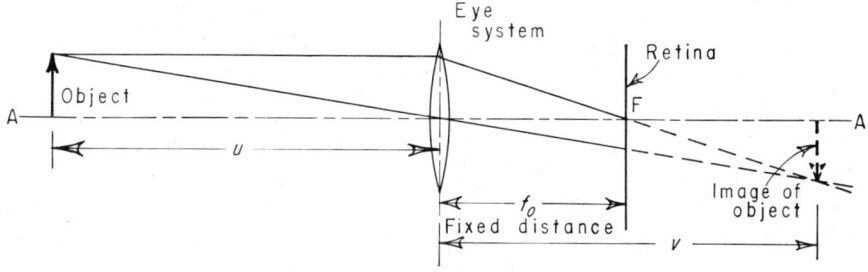

FIGURE 81. Moving a test object nearer an eye that is accommodated for a distant object causes relatively hyperopic imagery on the retina.

The distance from the equivalent "lens" to the retina is fixed anatomically. To clear the image of the near object on the retina—that is, to *focus* the image on the retina—the power of this lens—the dioptric system of the eye—must be increased. Physiologically it is accomplished by an increase in the power of the crystalline lens within the eye, a process described as *accommodation*. The power of the eye is accommodated to the changed optical vergence of the incident light produced when the object is brought near the eye (Fig. 82). Thus the power of the eye as a whole is increased, and the first focal point, $F$, of the eye no longer coincides with the retina but now lies in front of it, so that the image lies on the retina.

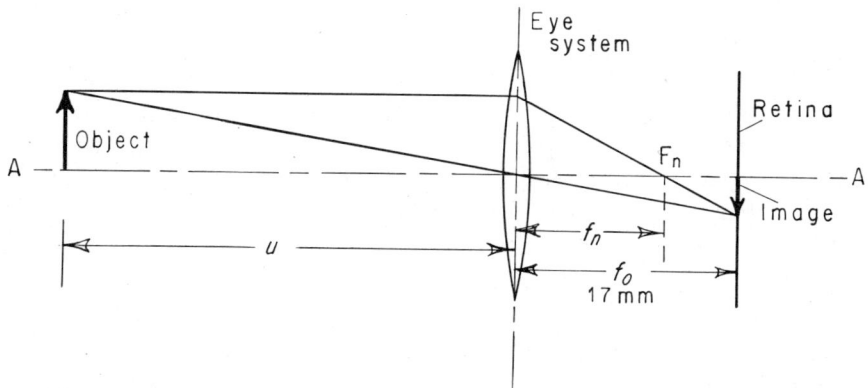

FIGURE 82. Increased dioptric power of the eye focuses the image of a near object on the retina.

Let $u$ be the object distance and $f_o$ the fixed image distance—that is, the distance from lens to retina. Then in terms of vergences and equation 19, in which $F_o = 1/f_o$ and $U = 1/u$, the accommodated dioptric power of the eye, $F_n$, is

$$F_o + U = F_n.$$

Thus the change in the power of the eye will be $F_n - F_o = U$. The dioptric power in the accommodation of the eye is equal to the dioptric vergence at the eye of the rays of light from the object at a given distance. An object at 40 cm requires a change of 2.5 diopters in the power of the optical system of the eye.

An eye which is not emmetropic is said to be *ametropic* or to possess a *refractive error*.

## B. Myopia

An eye is said to be myopic if the total refractive power is greater than that of the emmetropic eye having the same axial length. The vergence of the light transmitted is too great. The distance of the retina from the equivalent lens is then greater than the focal length of the unaccommodated eye. Hence the image of a distant object is formed in front of the retina, and interception of the rays by the retina results in a *blurred* image (Fig. 83). For the image to be sharply focused on the retina, either (1) the object must be brought closer to the eye, or (2) a minus ophthalmic lens must be placed before the eye to decrease the vergence of the rays incident on the cornea—that is, to correct the myopic error. These two methods will now be discussed.

For a given myopic eye, with relaxed accommodation, there is a particular point in front of the eye which is conjugate to the retina. The vergence of the light at the cornea is made sufficiently divergent by bringing the object up to this point. Thus the *far point* will be in front of the myopic eye and as such is the farthest distance at which the image is still sharply focused on the retina. Figure 84 indicates how we can diagram the lines representing rays to find the position of the far point of the myopic eye.

From the axial point, $C'$, on the retina, draw *any* line $C'P$ to the equivalent lens (eye system). From the intersection of this line with the focal plane at $F$, draw a line through the center of the lens (shown by an interrupted line in the diagram). This line

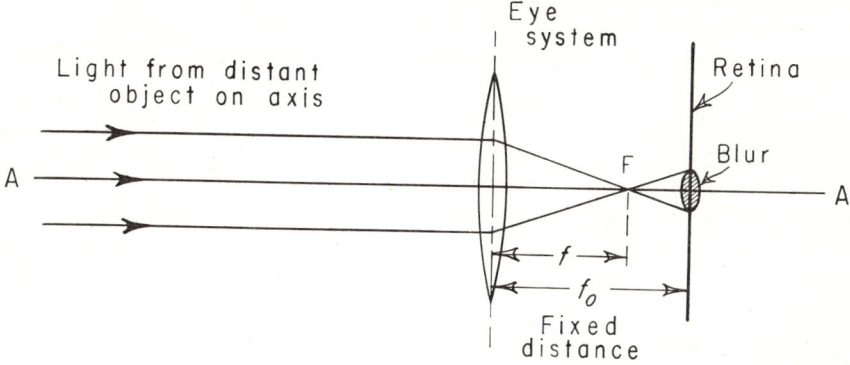

FIGURE 83. In the myopic eye the light rays from a distant object are focused in front of the retina.

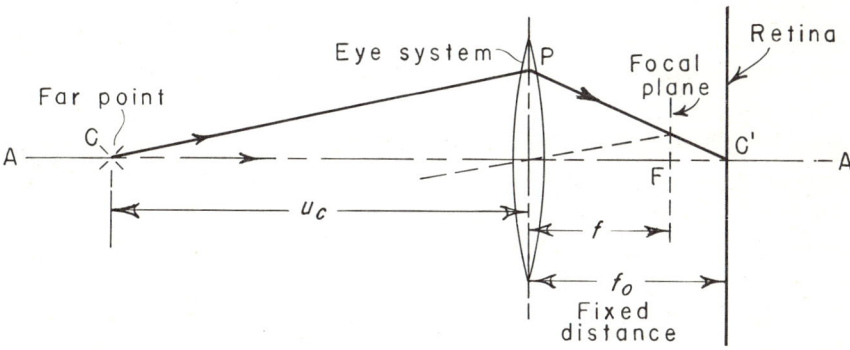

FIGURE 84. Method for determining diagrammatically the position of the point conjugate to the retina for a myopic eye.

representing a ray directed to the center of the lens is undeviated. From the point $P$ draw a line parallel to this undeviated line. The intersection of this newly drawn line with the axis locates point $C$, which by construction is the object point conjugate to the retinal point, $C'$. This method of diagramming follows from the rules (1) that all parallel rays incident on the lens come to focus at the same point in the focal plane, and (2) that the ray through the center of the lens is undeviated. Hence all rays originating at or directed through the point $C$ and incident on the eye are focused at the retinal point, $C'$. Myopia is commonly called nearsightedness, for the image can be made clear by bringing the object nearer to the eye.

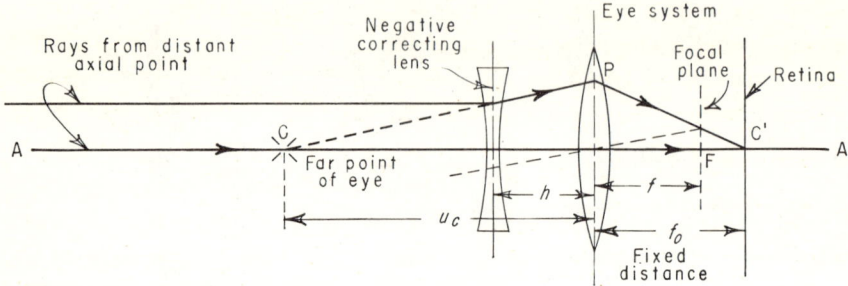

FIGURE 85. Correction of a myopic error with a diverging lens whose focal point coincides with the point (the far point) conjugate to the retina of the myopic eye.

To make the image of a *distant* object fall on the retina, a diverging correcting lens can be placed before the eye. This lens (Fig. 85) must be of such a power that its first focal point will coincide with the far point of the myopic eye. This is because the rays of light from a distant object, on passing through a diverging lens, emerge diverging as if originated from its first focal point. In the figure, this point is $C$, since the incident light is parallel to the axis, and all rays from point $C$ are brought to a focus at point $C'$ on the retina.

If the far point $C$ is the point conjugate to the retinal point $C'$, at a distance $u_c$ from the eye, and the correcting lens is placed at a distance $h$ from the eye, then the power of the correcting lens, $V_o$ diopters, will be

$$V_o = -\frac{1}{u_c - h}, \qquad \ldots (24)$$

which must be minus. In the clinic, of course, the far point of an eye is usually not determined directly; but rather that power, $V_o$, of the correcting lens is found which, when placed at a certain distance, $h$, before the eye, produces a clear image on the retina of the unaccommodated eye. With this power we can calculate the distance from the eye to the far point by solving for $u$, as $u_c = -[1 - V_o h]/V_o$ (meters). The true myopia (My), diopters, of the eye itself is the reciprocal of $u_c$, namely $My = 1/u_c = -V_o/(1 - V_o h)$, whence

$$-V_o = \frac{(My)}{1 - (My)h}. \qquad \ldots (25)$$

Thus the exact power of the diverging lens needed to correct a given myopia depends on the distance of the lens from the eye; this power must be (slightly) *greater* the farther the lens is from the myopic eye.

### C. Hyperopia

An unaccommodated eye which is hyperopic (farsighted) has a total refractive power less than that of the emmetropic eye of the same axial length; the vergence of the pencil of light transmitted through it is too weak. Hence the focal length of the accommodatively relaxed hyperopic eye is *greater* than the distance from the equivalent lens to the retina (the axial length): The first focal point of the dioptric system of the eye is behind the retina. Again, the rays from a distant object produce a blur on the retina (Fig. 86). The image can be brought into focus on the retina by increasing the vergence of the transmitted light in two ways: (1) accommodation of the eye to increase the total dioptric power of the eye as a whole, and (2) introduction of a converging lens before the eye to increase the vergence of the light rays incident to the eye.

In contrast to what is true of the myopic eye, the point conjugate to the retina for a hyperopic eye is *behind* the retina. This is demonstrated in Figure 87, which uses the same construction as Figure 84 in diagramming rays to determine the point conjugate to the retina. From the axial image point $C'$ on the retina,

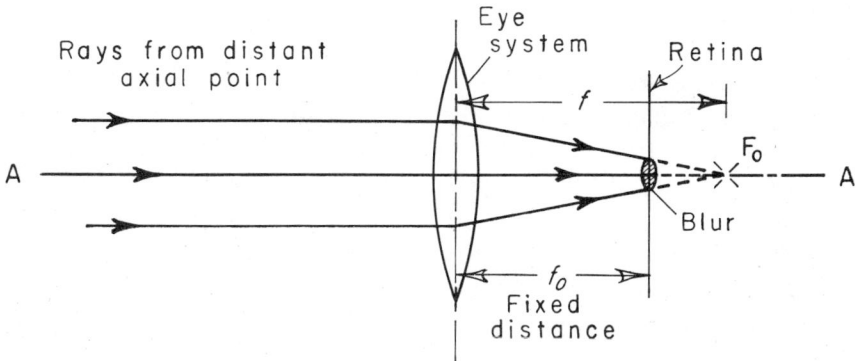

FIGURE 86. In the unaccommodated hyperopic eye the light rays from a distant object are focused behind the retina.

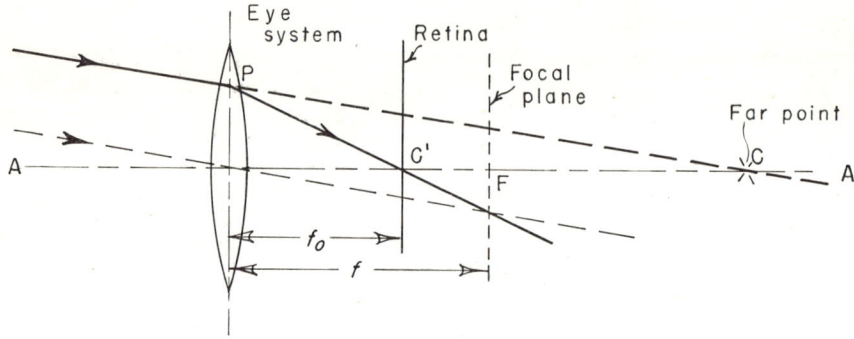

FIGURE 87. Method for determining diagrammatically the position of the point conjugate to the retina for an unaccommodated hyperopic eye.

*any* ray $C'P$ is drawn to the equivalent lens. From the intersection of this line with the principal focal plane at $F$, the focal point of the eye, a line (shown interrupted here) is drawn through the center of that lens. This line, representing a ray, is undeviated. Now from the point $P$ is drawn the line parallel to the interrupted line. The intersection of this line with the axis locates point $C$, conjugate to the retinal point $C'$. Thus, all rays directed to the point $C$ and incident on the eye are focused, on passing through the eye, at the retinal point $C'$. The point $C$ is therefore the *far point* of the hyperopic eye.

We cannot, of course, place a real object at this point, but rays from a distant object can be made to converge to this point by a *converging* lens of proper power placed before the eye. The first focal point of the correcting lens must coincide with the far point of the eye. Figure 88 illustrates the paths of rays from a distant object, converged by the correcting lens to the far point $C$ and then converged further by the hyperopic eye to focus at the retinal point $C'$.

Again, if the point conjugate to the retina, the far point $C$, is a distance $u_c$ from the eye, and the correcting lens is placed a distance $h$ from the eye, then the power of that correcting lens, $V_o$ diopters, should be

$$V_o = \frac{1}{u_c + h}.$$

In the clinic the far point itself is never actually determined, but

# Some Applications of Thin-Lens Theory to Visual Optics 109

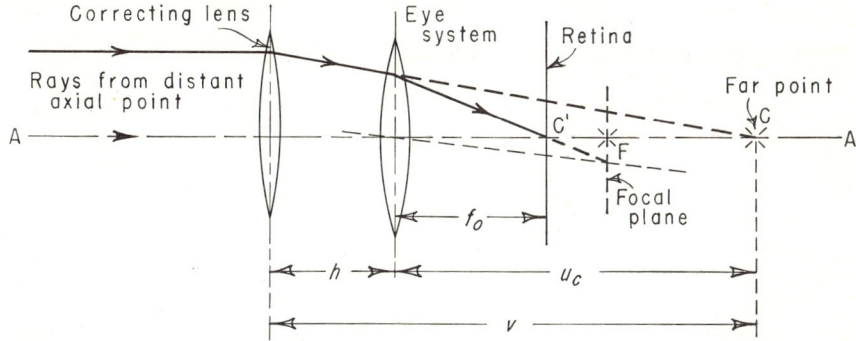

FIGURE 88. Correction of a hyperopic error with a converging lens whose focal point coincides with the point (the far point) conjugate to the retina of the unaccommodated hyperopic eye.

rather, what is determined is the power, $V_o$ diopters, of that correcting lens which, placed at a certain distance from the unaccommodated eye, produces a clear image on the retina. We can find the distance to the far point by solving the preceding equation for $u$, as $u_c = (1 - V_o h)/V_o$ (meters). The true hyperopia (Hy), if the eye is unaccommodated, is $(Hy) = V_o/(1 - V_o h)$, the reciprocal of $u_c$. Conversely, the power of the converging lens, $V_o$, necessary to correct a true hyperopia of (Hy) diopters is

$$V_o = \frac{(Hy)}{1 + (Hy)h}. \qquad \ldots (26)$$

Thus the power of the converging lens needed to correct a given hyperopia depends also upon the distance of the lens from the eye, the power being slightly *lower* the farther the lens is from the eye.

## D. Lens Power and Lens Position

A single formula can be used to relate the power to position of the lens. If we let (Am) diopters be the refractive error or ametropia of the eye and equal to the reciprocal of the distance to the far point (plus for hyperopia and minus for myopia), then the power, $V_o$, of the correcting lens at a distance, $h$ (meters), from the eye is given by

$$V_o = \frac{(Am)}{1 + (Am)h}, \qquad \ldots (27)$$

or, expressed in terms of the distance to the point conjugate to the retina, $u_c$ [and $u_c = 1/(Am)$],

$$V_o = \frac{1}{u_c + h}. \qquad \ldots (28)$$

Also, if we wish to find the position of the far point, knowing the power of the corrective lens $V_o$, we have

$$u_c = \frac{1 - V_o h}{V_o} = \frac{1}{V_o} - h. \qquad \ldots (29)$$

Again, $V_o$ is minus in myopia, plus in hyperopia. In either case, the point conjugate to the retina is a definite point for a given ametropia. The power of the correcting lens must be such that the first focal point of the lens coincides with the point conjugate to the retina, and consequently the required power of the lens is affected by the distance of the lens from the eye.

Very frequently we need to know the effect of a *change* in the distance from eye to lens upon the power of the lens to be prescribed. Suppose it is found that a −10.00-diopter lens placed 15 mm from a given myopic eye fully corrects the refractive error of that eye. The focal length of this lens is −0.100 meter or −100 mm, and the point conjugate to the retina of the myopic eye is then $u_c = -100 - 15 = -115$ mm in front of the eye (Figs. 85 and 89). Suppose, now, this same eye is to be corrected by a lens placed at $h' = 20$ mm from the eye instead of 15 mm. What should be the power of that lens? By inspection of Figure 89, the focal length of this lens would have to be $115 - 20 = 95$ mm. Hence, the power, $V_o$, of the lens would be $V_o' = -1/0.095 = -10.50$ diopters. Thus, increasing by 5 mm the distance of a minus lens from the eye has necessitated increasing the power of that lens by −0.50 diopter. Conversely, if the −10.00-diopter lens itself were moved from 15 to 20 mm from the eye, the myopia would be undercorrected by 0.50 diopter. The difference in the focal lengths of the lenses at the two positions equals the difference in distances of the lenses from the eye.

The somewhat inverse situation exists when hyperopia is corrected by a lens (Fig. 90), for now the point conjugate to the retina of the eye is behind the eye. Suppose it is found that a lens of power $V_o = +10.00$ diopters placed at a distance $h = 15$ mm in front of the eye fully corrects the hyperopia. The focal length

Some Applications of Thin-Lens Theory to Visual Optics        111

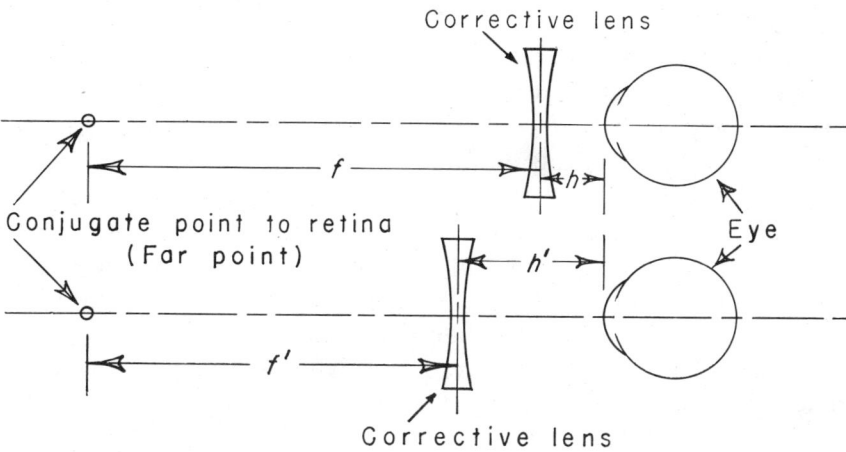

FIGURE 89. Influence of distance of a diverging lens from the eye upon the power of the lens required to correct the myopia.

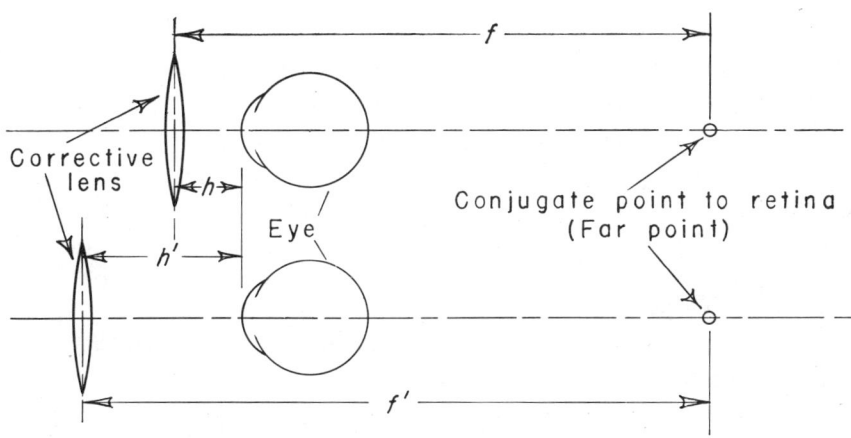

FIGURE 90. Influence of distance of a converging lens from the eye upon the power of the lens required to correct the hyperopia.

of this lens must be $f = 0.100$ meter or 100 mm. The point conjugate to the retina of the eye is $100 - 15 = 85$ mm behind the eye. What is the power of a lens that would correct the hyperopia of this eye when placed at $h' = 20$ mm instead of 15 mm? From inspection of Figure 90, the focal length of this lens would have to be $85 + 20 = 105$ mm; hence the power of this lens would have to be $V_o' = 1/0.105 = +9.52$ diopters. Thus, increasing the distance of a plus lens by 5 mm has necessitated decreasing the power by nearly 0.50 diopter. If the $+10.00$-diopter lens itself were moved to the 20-mm position, it would overcorrect the hyperopia by about 0.50 diopter.

A single formula can be worked out to cover this problem of the effect of the change in distance of a correcting lens upon the power required to correct the ametropia. For a given eye, the distance of the point conjugate to the retina remains constant; so if $V_o$ is the power required when the lens is placed $h$ meters from the eye, and $V_o'$ is the power required when the lens is placed $h'$ meters from the eye, from relationship 29 we have

$$u_c = \frac{1 - V_o h}{V_o} = \frac{1 - V_o' h'}{V_o'},$$

whence we find for the power of the corrected lens, $V_o'$, at the new position $h'$,

$$V_o' = \frac{V_o}{1 + V_o(h' - h)}. \qquad \ldots (30a)$$

We can write for the displacement of the lens $(h' - h) = (\Delta h)$ meters. If a converging lens is moved away from the hyperopic eye [$(V_o)$ and $(\Delta h)$ both being plus], the power of the lens must be reduced to correct the hyperopia. Similarly, a converging lens of a given power, $V_o$, thus moved would overcorrect the hyperopia. When contact lenses are used, the power of the converging-lens correction must be markedly increased because $h'$ now is much less than $h$. The reverse situation obtains in the case of diverging lenses.

Now if $V_o$ is not too large and the shift of lens position, $(\Delta h) = (h' - h)$, is not too great, the power of the lens required after the shift of position can be written (from a series expansion) in the approximate form

$$V_o' = V_o[1 - V_o(\Delta h)],$$

or

$$V_o' = V_o - V_o^2(\Delta h). \qquad \ldots (30b)$$

The change in power $(\Delta V_o) = (V_o' - V_o)$ would then be given by

$$(\Delta V_o) = -V_o^2(\Delta h), \qquad \ldots (30c)$$

$(\Delta h)$ being in meters. This is the well-known formula to be found in every textbook on ophthalmic optics. However, it is only an approximate formula and is adequate for general use only if the lens powers and the lens displacements are not high. If we express ($\Delta h$) in millimeters, then the change in dioptric power required for a shift of the lens away from the eye ($\Delta h$) mm is

$$(\Delta V) \text{ (diopters)} = -V_o^2(\Delta h)/1000.\phantom{xx}^*$$

In the following table are given approximate magnitudes for the *change* of power required to correct the ametropia for a 1-mm displacement of lens.

| Lens power (diopters) | 2.5 | 5 | 7.5 | 10† | 15† | 20† |
|---|---|---|---|---|---|---|
| ($\Delta V_o$) (diopters) | 0.006 | 0.025 | 0.056 | 0.100 | 0.225 | 0.400 |

†For these lens powers and larger displacements the formula is only approximate.

This table shows that changes of even 5 mm in the distance of the lens from the eye result in little appreciable change in the required power of the lens unless the lens power is greater than 7 diopters.

### E. Effective Power

The term *effective power*, as applied to the power of a given ophthalmic lens at a certain position before the eye, refers to the optical effect of the lens as if it were at some other position, usually a standard or reference position, before the eye. For example, trial-case ophthalmic lenses usually placed at a certain distance from the eye, say 13 mm, are used in determination of the power of the corrective lens for the refractive error of an eye. If, for one reason or another, the lens is later placed at a different distance, then the question arises as to what is the *effective power* of the lens relative to the reference distance, 13 mm, so that the lens could be prescribed as if it were there.

By use of the same relationships derived in the previous section, if a lens of power $V_o$ diopters is moved a distance ($\Delta h$) meters from the eye, the change in the effective power ($\Delta V_o$) of that lens referred to its original position would be

$$(\Delta V_o) = +V_o^2(\Delta h).$$

The effective power itself would then be

---

*In certain textbooks this formula is written $(\Delta D) = -KsD^2$, in which $K = 1/1000$. $s$ is the shift of lens position in mm ($=\Delta h$), and $D$ is the power of the lens in diopters ($=V_o$).

$$V_o' = V_o + (\Delta V_o).$$

*Example:* A +8.00-diopter lens at 16 mm from the eye fully corrects the hyperopia of the eye. What is the effective power of this lens placed at a distance of 12 mm from the eye? Given, then, are $V_o = +8.00$, and $(\Delta h) = h' - h = 12 - 16 = -4$ mm, or $(\Delta h) = -0.004$ meter. Hence the effective power of the lens is $V_o' = +8.00 - (8.00)^2 (0.004) = +7.75$ diopters. This means that a +8.00-diopter lens which would fully correct the given hyperopia at a distance of 16 mm has an effective power of only 7.75 diopters when placed at a distance of 12 mm from the eye.

The effective power of a plus lens moved away from the eye is increased, while that of a minus lens is decreased. The reverse is true when the lens is moved toward the eye.

## EXERCISES

1. A +2.50-diopter lens at 15 mm from the eye is needed to correct a completely presbyopic eye so that print at 40 cm from the lens can be read. What is the angular magnification of the print seen through the lens? What is the per cent angular magnification?
2. A converging lens of +15.00 diopters is to be used as a simple microscopic lens. What is the angular magnification produced compared to print at the usual reading distance of 40 cm?
3. If print which is at 40 cm from the eyes requires a visual acuity of 20/20 in order to be read by a normal person (details subtend 1 minute of arc), and if this same print is barely legible to the patient with the microscopic lens described above, what is the probable visual acuity of this patient without the lens?
4. Suppose a so-called microscopic lens is needed for a patient with 20/200 vision so that he can read 20/200-size print (normally seen at 40 cm). What should be the power of the lens? If the lens is held 15 mm from the eye, how close to the eye does print have to be held?
5. Determine the positions of the far point (the point conjugate to the retina) for eyes corrected by lenses at a distance of 15 mm and having powers of $+5$, $-8$, $+10$, and $-15$ diopters.
6. A hyperopic eye is corrected by a +5.00-diopter lens at a distance of 15 mm. Determine the dioptric power of an ophthalmic lens (considered a thin lens) which placed at 6.5 cm also corrects this eye. Calculate by rigorous and approximate methods. What is your conclusion?
7. A corrective lens of −10.00 diopters at 20 mm corrects a myopic eye. What would be the effective power of this lens if it were 12 mm from the eye?
8. A +12.50-diopter lens at a distance of 12 mm corrects an aphakic eye. If a corrective lens were placed in contact with the cornea, what

would its power have to be? Is formula 30b sufficiently accurate for lens powers of this magnitude?

9. A +12.50-diopter lens corrects an aphakic eye at a distance of 15 mm. At what distance from the eye must a +3.00-diopter lens be held in order to also correct the aphakia?

## REFERENCE

1. FONDA, GERALD: *Management of the Patient with Subnormal Vision.* St. Louis, Mosby, 1965.

*Chapter VII*

# IMAGE FORMATION BY COMBINATIONS OF THIN LENSES

The discussion in this chapter will eventually lead to the optical theory of thick lenses, for the reason that the two surfaces of a thick lens can be considered as a combination of two thin lenses when appropriate corrections are made for the thickness of the glass.

## 1. BASIC CONSIDERATIONS

The thin lenses of the combinations to be considered here are assumed to be coaxial or centered—that is, so arranged that the optic axes of all the lenses in the combination coincide with each other. The image formed by the first lens of any sequence of such lenses can be considered as the object for the second lens, and so on. The relationship between the image-object distances, as already developed for thin lenses, can be used successively in proceeding from one lens to the next in the sequence.

Consider more explicitly the case of only two thin lenses separated by a distance $s$ (Fig. 91). The relative positions of the object and image can be diagrammed in the usual manner. For simplicity in this case, let each of the two be a converging lens. This stipulation in no way limits the generality of the results. If, for the first lens (1), $u_1$ and $v_1$ represent the object and image distances (meters), respectively, as before, and $F_1$ is the power in diopters, we have

$$\frac{1}{v_1} + \frac{1}{u_1} = F_1. \qquad \ldots (31)$$

Now, if the image formed by the first lens is the object for the second lens, then the distance, $u_2$, of this object from the second lens (2) is

# Image Formation by Combinations of Thin Lenses

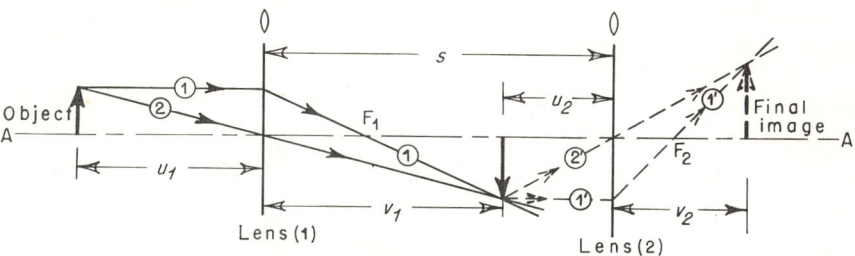

FIGURE 91. Method of diagramming specific rays through a combination of two separated thin lenses to locate the image of an object.

$$u_2 = s - v_1. \quad \ldots (32)$$

The image-object relationship for the second lens, the dioptric power of which is $F_2$, then is

$$\frac{1}{v_2} + \frac{1}{u_2} = F_2. \quad \ldots (33)$$

If we manipulate equations 31, 32, and 33 algebraically to eliminate $u_2$ and $v_1$, we can obtain the general relationship

$$u_1[1 - F_1 s] + v_2[1 - F_2 s] - u_1 v_2[F_1 + F_2 - F_1 F_2 s] + s = 0. \quad \ldots (34)$$

This equation is to be compared to the similar relationship between image and object distances for a single thin lens—namely $u + v - Fuv = 0$, obtained from $1/v + 1/u = F$. From equation 34 the value for the distance of the final image from the second lens can be found by solving for $v_2$:

$$v_2 = \frac{u_1(1 - F_1 s) + s}{u_1(F_1 + F_2 - F_1 F_2 s) - (1 - F_2 s)}. \quad \ldots (35)$$

This is a relatively complicated formula for general calculation.

*Note:* If the separation of these lenses is very small or approaches zero—that is, if the lenses are in contact—then from equation 34 we have

$$u_1 + v_2 - u_1 v_2 (F_1 + F_2) = 0.$$

Dividing through by $(u_1 v_2)$, to give vergences, and setting $u_1 = u$ and $v_2 = v$, we have

$$V + U = (F_1 + F_2).$$

Comparison with equation 19 shows that the power of two lenses in contact is equal to the sums of the individual powers.

The image distance, $v_2$, from the second lens will be the focal distance of the combination if, in equation 35, $u_1$ is made very large. Then, $1/u_1 \to 0$, whence

$$v_2\Big]_{u_1 \to \infty} = \frac{1 - F_1 s}{F_1 + F_2 - F_1 F_2 s}. \qquad \ldots (36)$$

This focal distance is sometimes called the *back* or *vertex* focal length.

The magnification of the final image compared to the object would follow from $M_1 = I_1/O$ and $M_2 = I_2/I_1$; hence

$$M = \frac{I}{O} = M_1 M_2 = \left[\frac{v_1}{u_1}\right]\left[\frac{v_2}{u_2}\right]. \qquad \ldots (37)$$

## 2. PRINCIPAL POINTS AND PRINCIPAL PLANES

In any centered optical system, irrespective of the number of lenses, there exist on the axis two unique points whose use as reference points simplifies the relationship between the object and image distances of the system. They are called *principal points*. Their positions can be located experimentally on the lens bench and for simple lens combinations can be determined by calculation. If these points can be found or can be determined by calculation, and then the object distance is measured from one and the image distance from the other, the simple object and image formula for thin lenses can be used—namely,

$$\frac{1}{v} + \frac{1}{u} = F,$$

in which $F$ is the true power of the lens combination. Imaginary planes erected perpendicularly to the axis of the combination through these points are called *principal planes*, and these are useful for diagramming the course of rays through the combination.

In Figure 92, two thin lenses are shown separated by a distance $s$, as before. The object distance $u$ is measured to the first principal point (plane), which is a distance $a$ from the first lens, while the image distance $v$ is measured from the second principal point (plane), which is a distance $b$ from the second lens.

The formulas by which $a$ and $b$ can be derived in a two-lens system are as follows. Referring to Figures 91 and 92, the object distance $u_1$ measured from the first lens is

$$u_1 = u - a,$$

and, similarly, the image distance $v_2$ measured from the second lens is

$$v_2 = v + b.$$

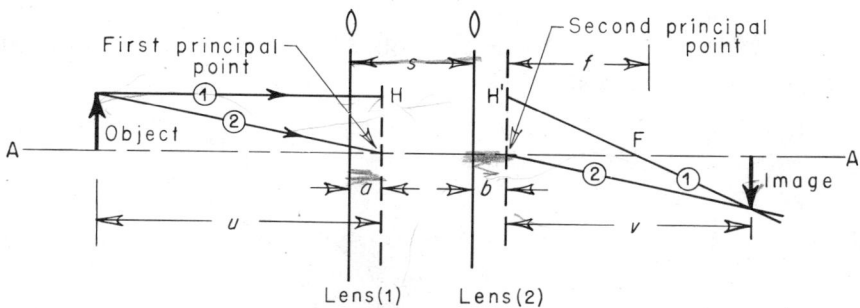

FIGURE 92. The use of principal points simplifies the relationship between object and image distances in a combination of two lenses.

Substituting these equivalents for $u_1$ and $v_2$ in the general equation 34, and simplifying, we obtain

$$\overset{①}{[u + v - uvF]} - \overset{②}{[v(F_2 s - aF)} + \overset{③}{u(F_1 s + bF)]}$$
$$\overset{④}{- [a - b - F_1 a s + b F_2 s - Fab - s]} = 0,$$

in which $F = F_1 + F_2 - F_1 F_2 s$, the power of the combination.

The term ① will be zero if both terms ② and ③ are zero. These latter two terms can be made zero if we set

$$a = \frac{F_2}{F} s, \text{ and } b = -\frac{F_1}{F} s. \qquad \ldots (38)$$

If we substitute the values for $a$ and $b$ in term ④, then term ④ is also equal to zero. These two formulas give a method of calculating the positions of the principal points of a combination of two thin lenses. [The distance from the first lens to the *first principal point* is $a$, and the distance from the second lens to the *second principal point* is $b$. If either $a$ or $b$ is minus, it is to be measured to the left of the lens.]

If, now, we measure the object distance $u$ from the first principal plane, and the final image distance $v$ from the second principal plane,

$$u + v - uvF = 0,$$

whence

$$\frac{1}{v} + \frac{1}{u} = F, \qquad \ldots (39)$$

which is our simple relationship of object-image distances. Thus, by using the device of principal points, we can reduce a complex optical system to a simple one.

The quantity

$$F = F_1 + F_2 - F_1 F_2 s \qquad \ldots (40)$$

is the *true* power of the system. The true focal length of the combination is $f = 1/F$; this distance is measured, of course, from the second principal point, since when $u \to \infty$

$$\frac{1}{v}\Big]_{u \to \infty} = \frac{1}{f} = F.$$

The anterior focal length also is $f$, but measured from the first principal point.

The magnification of the final image produced by the lens combination relative to the object also is

$$M = \frac{I}{O} = \frac{-v}{u}.$$

This relationship must mean that a ray directed to the first principal point will leave the combination as if from the second principal point, the incident and emerging rays making the same angle of inclination with the axis. This fact makes it possible to determine experimentally the positions of the two principal points of any centered lens system.

It can be shown that the principal points themselves are conjugate points. We use equation 34 for the object and image distances measured from the first and second lenses of a two-lens combination. We substitute for the object distance $u_1 = -a = -F_2 s/F$ (minus because in Figure 92 $a$ is drawn to the right of the first lens), and then solve for $v_2$, whence we find $v_2 = F_1 s/F = b$. It can also be shown that the magnification of an image for an object at the first principal point is one; that is, the image and object are the same size.

With these concepts in mind, it is easy to diagram the rays necessary to relate the positions of the object and image. In Figures 92 and 93 the positions of the principal planes are known. Then the line representing a ray (1) is drawn parallel to the axis. This line intersects the first principal plane at a given height above the axis and emerges from the system at a point at the same

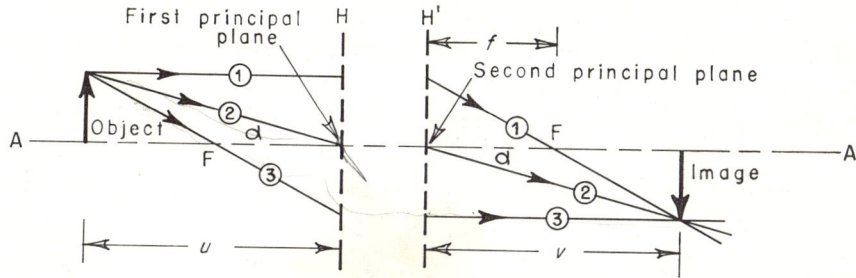

FIGURE 93. Method of diagramming the course of specific rays to locate the image of an object by use of the principal planes of the optical system.

height on the second principal plane (because the magnification of image and object at the principal points is unity). This ray then proceeds through the first focal point of the combination. As for the thin lens, another line representing a ray (2) can be directed to the first principal point; this line leaves the lens combination at the second principal point, making the same angle to the axis. The intersection of these lines (1 and 2) determines diagrammatically the position of the image. Similarly, a third line representing a ray (3) can be directed through the second or anterior focal point to intersect the first principal plane at a certain height; this line leaves the system at a point at the same height on the second principal plane and parallel to the axis. It also passes through the same image point. In simple terms, the construction used is as if the thin lens had been split and the two halves moved to the principal points.

The positions of the principal points may vary greatly with the powers of the components making up the combinations of thin lenses. Combinations of more than two thin lenses can also be treated in pairs, and these pairs combined, and the positions of final principal points determined thus. However, in actuality it is only for simple lens combinations, such as ophthalmic lenses, that the positions of the principal points can be easily determined by calculation. In more complicated lens systems, such as photographic lenses, the locations of the principal points (although they can be calculated if the dimensions of the component lenses are known) are best determined experimentally.

The experimental method of locating the principal points of an optical system rests upon the fact that a ray directed to the first principal point emerges from the system as if from the second principal point. When such a system is rotated or oscillated through a small angle, about an axis through the second principal point, the image formed by the system will not move on an image screen. The procedure is to place the lens on a lens bench in a special holder that permits us to move the lens fore and aft over a pivot point and to find by trial and error that position of the lens (with respect to this pivot point) for which the image remains stationary.

The use of principal points and the concept of principal planes reduces a complex optical system to a simple one in which the familiar relationships between the positions of the object and the image and their magnification apply.

### 3. TELESCOPIC SYSTEMS

A telescopic optical system is one in which the image of an infinitely distant object is also at an infinite distance. The ordinary telescope, field glasses, and opera glasses are of this type. Parallel rays of light incident to the system emerge from the system parallel. Such a system, not changing the vergence of the incident light, therefore has zero power and is *afocal*. However, there is an angular magnification of the image which can be appreciated by the eye looking through the system.

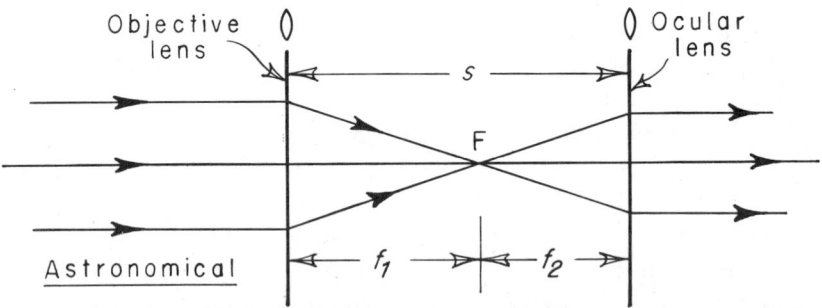

FIGURE 94. Relationship between the focal lengths of two plus lenses which form an afocal (telescopic) lens system of the astronomical type.

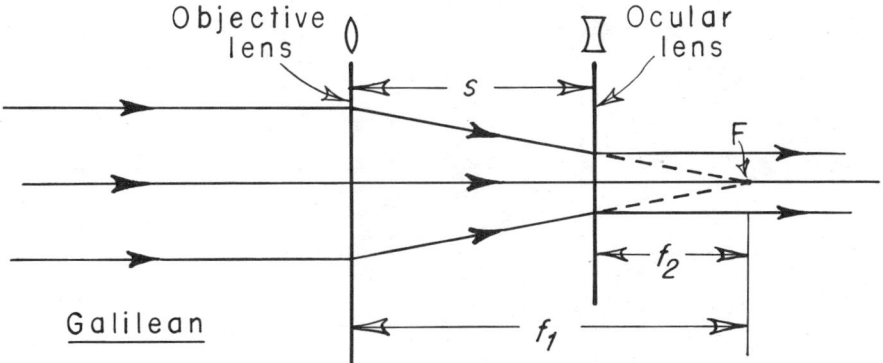

FIGURE 95. Relationship between the focal lengths of a plus and a minus lens which form an afocal (telescopic) lens system of the Galilean type.

For a two-element combination the relationship between the powers of the component lenses (equation 40) must be

$$F = F_1 + F_2 - F_1 F_2 s = 0. \qquad \ldots (41)$$

If we substitute for the powers in equation 41 their reciprocals—namely, the focal lengths $f_1$ and $f_2$—and clear the equation of fractions by multiplying through by $f_1 f_2$, we obtain

$$f_2 + f_1 - s = 0 \text{ or } f_1 + f_2 = s. \qquad \ldots 42$$

This relationship tells us how far the two lens elements must be separated to make the system telescopic. Theoretically there is a particular separation for any two lenses (except when both are minus) which makes their combination afocal. If the powers of the two lens elements are both plus, both focal lengths are plus; then the separation must be equal to the sum of the focal lengths. Such a system is commonly called an *astronomical* telescope, and it is schematically illustrated in Figures 94 and 96.

If the second of the two elements is a diverging (or negative) lens, the combination is then commonly called a *Galilean* system. In this case,

$$f_1 - f_2 = s.$$

Such a system is illustrated schematically in Figures 95 and 97. This system has the advantages of being shorter and providing an erect image. Most opera glasses are of the Galilean type.

In these particular diagrams the smaller separation between

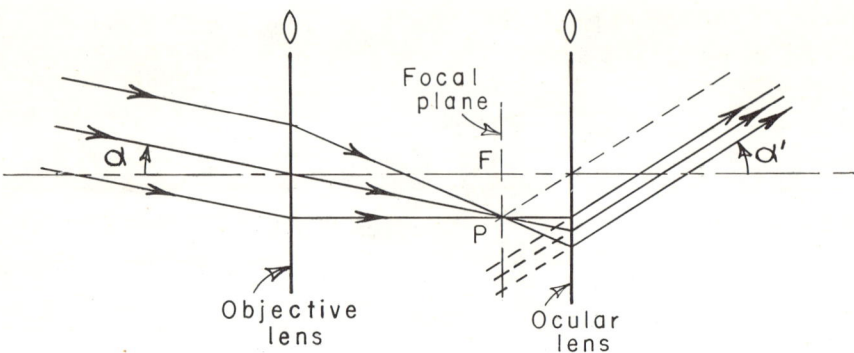

FIGURE 96. Method of tracing rays through the astronomical type of telescope to show inverted imagery and angular magnification.

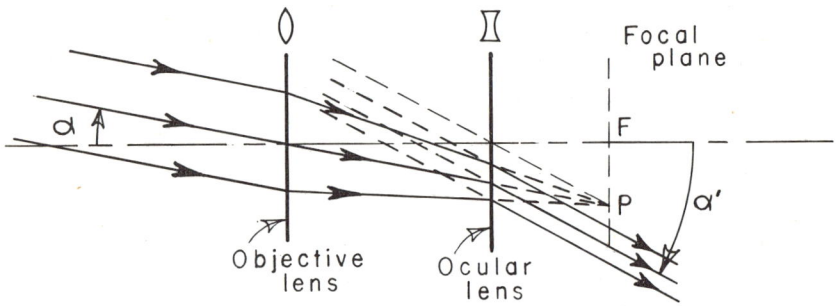

FIGURE 97. Method of tracing rays through the Galilean type of telescope or afocal system to show erect imagery and angular magnification.

emerging rays compared to the corresponding separation between the incident rays does not mean that the image size is reduced but rather that the light energy incident to the first element—the *objective* lens—is more concentrated when it leaves the second element—the *ocular* lens.

We may diagram the rays to show the angular magnification of these systems as illustrated in Figures 96 and 97, for the *astronomical* and *Galilean* telescopes, respectively. In these, parallel rays are drawn to the objective lens, the central line (chief ray) being directed to the lens center and making an angle $\alpha$ with the axis. This chief ray is undeviated and meets the focal plane of the objective at the image point, $P$. Since this plane also is the focal plane of the ocular, the chief ray from the point P through the center of the ocular is undeviated; and since all the rays corresponding to the incident rays emerge parallel from the ocular, they must be parallel to this line. Thus the rays from the point P, incident to the ocular lens, emerge parallel to the chief ray. The angle $\alpha'$ between the emerging rays and the axis is greater than the angle of incidence, $\alpha$; thus the angular size is increased. In the astronomical system the image is *inverted*. In the Galilean system the image is *erect*.

A distant object viewed through one of these telescopic systems, therefore, appears magnified. The angular magnification can be found readily if the powers of the objective ($F_1$) and the ocular ($F_2$) lenses and their separation $s$ are known. This angular magnification is $A = \tan \alpha'/\tan \alpha$ (or $A = \alpha'/\alpha$ if the angles are small). The image of the object (Figs. 96 and 97) formed by the objective lens is $FP$, and this image serves as the object for the ocular lens. We have by inspection: $\tan \alpha = FP/f_1$ and $\tan \alpha' = -FP/f_2$. Hence

$$A = -\frac{f_1}{f_2} = -\frac{F_2}{F_1} = \frac{1}{1 - F_1 s}. \qquad \ldots (43)$$

The focal length $f_2$ is also equal to $s - f_1$; hence $F_2 = -1/(1 - F_1 s)$.

There is another class of system similar to that of the telescope, called *terrascopic*. In this, the powers of the elements and their separation are so selected that the image formed of a near object will coincide with the position of that object. A magnification also results. In the notation used in the preceding material, $v_2 + s = -u_1$. This also is a useful system.

## EXERCISES

1. A +5.00-diopter thin lens is placed 12.5 mm in front of a +8.00-diopter thin lens.
    (a) What is the true power of the combination?
    (b) What is the true focal length?
    (c) Calculate the positions of the two principal points.
    (d) What is the distance of the posterior focal point from the second lens? Of the anterior focal point from the first lens?
    (e) An object is placed 40 cm in front of the first lens. How far is the image from the second lens?
    (f) What is the magnification?

2. If a −4.00-diopter lens replaces the +8.00-diopter lens used in the previous example, what are the answers to problems (a) through (f) above?
3. The powers of the front and back thin lenses of a Galilean-type telescope are +10.00 diopters and −12.50 diopters, respectively. What must the separation of the two lenses be if the system is to be afocal? What is the angular magnification? Would such a doublet have any value in ophthalmic optics?
4. Illustrate by diagram the rays through a lens combination examined on the lens bench, showing that if the combination is turned about an axis through the second principal point, the image of a distant object point remains stationary.

*Chapter VIII*

# IMAGE FORMATION BY REFRACTION FROM SPHERICAL SURFACES

THE refraction of light by curved surfaces and the optical theory of image formation by thick lenses can now be discussed.

Any spherical surface which separates two transparent media having different indices of refraction acts as a lens, in that its refraction of the incident light from an object forms an image. This discussion will be concerned with the relationships between the image and object distances; it will show that, under the simplest conditions, these relationships are essentially of the same general form as those for ideal thin lenses. The derivation of the simple formulas that provide the basis for the general discussion of ophthalmic lenses follows from the diagrams drawn by line constructions which obey the laws of geometric optics.

## I. REFRACTION BY SPHERICAL SURFACES

Consider a spherical surface of radius $R$, separating two transparent optical media having indices of refraction $n_1$ and $n_2$, respectively (Fig. 98). Let $n_2$ be greater than $n_1$, as would be the case when light passes from air to glass. In front of the surface is an object point, $S$. The light from this source passes through the spherical surface and is brought to a focus at an image, $S'$. Only two rays are necessary to locate that image. A ray from $S$ directed to the center of curvature of the surface passes through the surface undeviated (because the angle of incidence is zero, therefore the angle of refraction is zero). This ray defines an axis of the surface. Consider now a second ray from $S$. Suppose it is inclined at an angle, $\alpha$, from the first or axial ray. This second ray strikes the spherical surface at a point, $P$, at a height, $d$, above

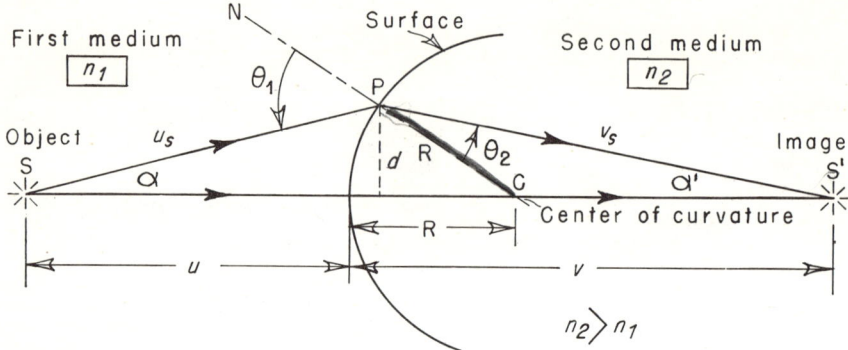

FIGURE 98. Method of diagramming specific rays refracted by a convex spherical surface to locate the image of a point object. (The figure is exaggerated to make identification of its features easier.)

the axial ray. The angle of incidence to the surface is $\theta_1$, an angle formed by the incident ray and the line perpendicular to the surface at $P$ (a line from the center of curvature, $C$, extended through $P$). The angle of refraction is designated by $\theta_2$. The refracted ray intersects the axial ray at $S'$, which is the position of the image, as determined by these two selected rays. As usual, the distance from the object to the surface is designated as $u$, and that from the image to the surface as $v$. To apply Snell's law of refraction and thereby to find the connection between object and image distances, $u$ and $v$, the relationships between the sides and angles of certain triangles in the figure are used.

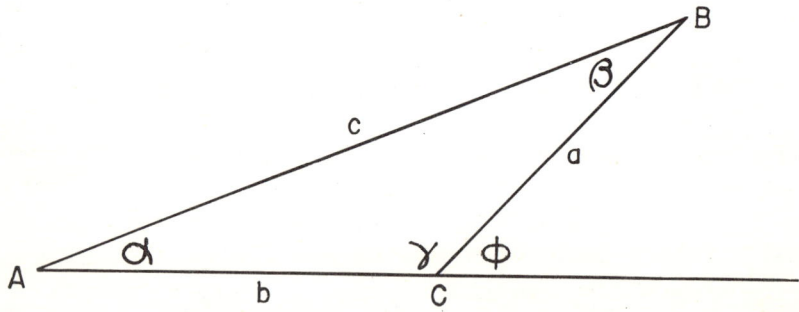

FIGURE 99. Triangle used to illustrate the law of sines, which relates the sides to the opposite angles.

*Note:* In trigonometry, the sides of any given triangle are related to the opposite angles by a law called the "law of sines." For the illustrative triangle in Figure 99, this law states, $\dfrac{a}{\sin \alpha} = \dfrac{b}{\sin \beta} = \dfrac{c}{\sin \gamma}$. Other well-known angular relationships of this triangle are $\alpha + \beta + \gamma = 180°$ (the sum of the angles of any triangle is equal to two right angles), and the external angle $\phi = \beta + \alpha$. Thus, $\phi + \gamma = 180°$, and $\sin \phi = \sin \gamma$. We shall use these laws in the solution of the present problem.

For the triangle *SCP* (Fig. 98), the angle of incidence, $\theta_1$, is an external angle, so we can write

$$\frac{R}{\sin \alpha} = \frac{R + u}{\sin \theta_1}. \qquad \ldots (44)$$

Similarly, in the triangle *CPS'*, where $\theta_2$ is an internal angle, we have

$$\frac{R}{\sin \alpha'} = \frac{v - R}{\sin \theta_2}. \qquad \ldots (45)$$

Now, by Snell's law

$$n_1 \sin \theta_1 = n_2 \sin \theta_2.$$

Solving equation 44 for $\sin \theta_1$ and equation 45 for $\sin \theta_2$, and substituting these values in the equation for Snell's law, we obtain on simplification

$$\frac{n_1(R + u)}{n_2(v - R)} = \frac{\sin \alpha'}{\sin \alpha}. \qquad \ldots (46)$$

This equation points up the fact that the relationship between the object and image distances ($u$ and $v$, respectively) varies with the angle $\alpha$. This is to say that for a given object distance the distance of the image, $S'$, varies slightly with the height of $P$, where the ray intersects the surface above the axis. In fact, equation 46 is a quantitative expression of *spherical aberration*.

The equation is difficult to reduce to a simpler but still exact expression. Referring again to Figure 98, however, let the distances $SP = u_s$ and $PS' = v_s$ represent the slant distances of the object and image from the point of incidence on the surface. Then we have

$$\sin \alpha = d/u_s \text{ and } \sin \alpha' = d/v_s.$$

The ratio $\sin \alpha'/\sin \alpha$ is then equal to $u_s/v_s$. Thus equation 46 becomes

$$\frac{n_1(R + u)}{n_2(v - R)} = \frac{u_s}{v_s}. \qquad \ldots (47)$$

To simplify the problem further, it is necessary to restrict the discussion to rays that lie close to the axis, the so-called *paraxial* rays, so that $\alpha$ and $\alpha'$ will then be small, and $u_s \to u$ and $v_s \to v$. Under that condition, relationship 47 reduces to

$$n_1(R + u)v = n_2(v - R)u,$$

which becomes, on simplification,

$$\frac{n_2}{v} + \frac{n_1}{u} = \frac{n_2 - n_1}{R} \cdot = D \qquad \ldots (48)$$

This is the fundamental relationship between the object and the image distances for a spherical surface and is the basic relationship between the object and image distances for all thick-lens theory. A spherical surface convex toward the incident light and separating two optical media, the index of the second being greater than that of the first, acts as a converging thin lens.

The quantity $(n_2 - n_1)/R$ is called the *dioptric power* $(D)$ of the surface. This power is equal to the difference in the index of refraction between the second and the first media divided by the radius of curvature of the surface. If $R$ is given in meters, $D$ will be in diopters. From equation 48, if $u$, the object distance, is very great, so that the incident light is parallel, $v$ is identical to the focal length $f$ in the second medium having index of refraction $n_2$—namely, $f = n_2/D$. Since the index of refraction of glass is about 1.5, a spherical surface separating air and glass has a focal length in the glass approximately three times the radius of curvature.

If the spherical surface is concave toward the incident light (Fig. 100), and again the index of refraction of the first medium is less than that of the second ($n_2 > n_1$), then the surface acts as a diverging thin lens. The image is virtual and on the same side of the surface as the object or source of the incident light. The radius of curvature $R$ is here taken *minus* (center of curvature to the left of the surface); hence the dioptric power of the surface is minus. With this change the same equation 48 holds.

Equation 48 is an important one. The difference between it and the equation for a thin lens lies only in the fact that the indices of refraction are involved. Of course the image, even when virtual, insofar as these equations are concerned, always lies in the media of index $n_2$.

*Image Formation by Refraction from Spherical Surfaces* 131

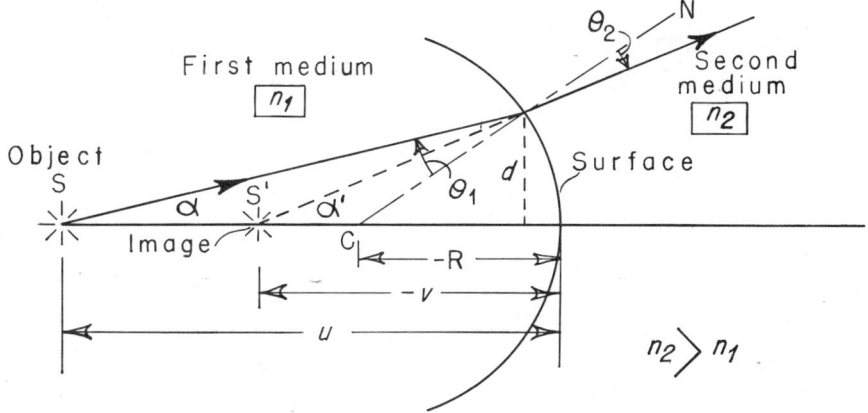

FIGURE 100. Method of diagramming specific rays refracted by a concave spherical surface to locate the image of a point object.

Whether a given surface has a plus (converging) or a minus (diverging) dioptric power depends on the difference in the indices of refraction on the two sides of the surface and whether the surface is concave or convex toward the incident light (see Fig. 101). In each case we should keep in mind the equation for dioptric power of the surface—namely, $D = (n_2 - n_1)/R$.

*Example 1:* The radius of curvature of the cornea of the human eye is about 7.7 mm ($R = +0.0077$ meter); the index of refraction for the corneal substance ($n_2$) is 1.376; and the index of refraction for air ($n_1$) is 1. Hence the dioptric power of the front surface of the cornea is

$$D_1 = \frac{1.376 - 1}{0.0077} = \frac{0.376}{0.0077} = 48.8 \text{ diopters.}$$

*Example 2:* The radius of curvature of the posterior surface of the cornea is about 6.8 mm. The index of refraction of the aqueous humor is 1.336 (nearly the same as that of water). Thus, $R = +0.0068$, $n_1 = 1.376$, and $n_2 = 1.336$. The dioptric power of the posterior surface is therefore

$$D_2 = \frac{+1.336 - 1.376}{+0.0068} = \frac{-0.040}{0.0068} = -5.9 \text{ diopters.}$$

*Example 3:* The surface powers of ophthalmic lenses are also defined in the same way. In the following table, where the index of glass ($n$) is 1.532, representative values are given for the powers and the corresponding radii of curvature of certain selected ophthalmic lens surfaces.

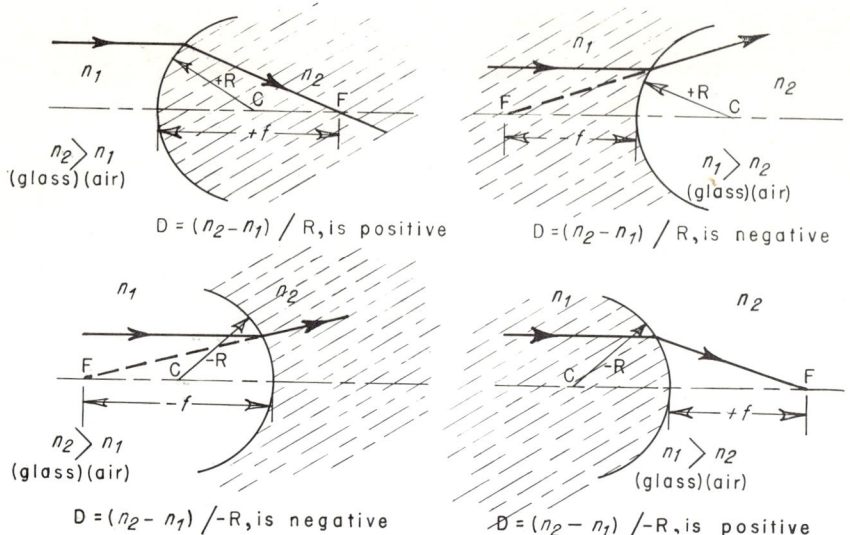

FIGURE 101. Representative cases of the sign of the dioptric power for concave and convex spherical surfaces and for different indices of refraction of the media on the two sides of the surface.

| Dioptric power, $D$ (diopters) | 0 (plano) | 2 | 6 | 10 | 15 |
|---|---|---|---|---|---|
| Radius of curvature, $R$ (cm) | ∞ (infinite) | 26 | 8.7 | 5.2 | 3.5 |

For a series of lenses, all having one surface with the same curvature, the power of that surface is often referred to as the *base curve* of the lens series.

## 2. MAGNIFICATION OF THE IMAGE FROM SINGLE REFRACTING SURFACES

Consider now the image formed by the surface for an object of finite size $AB$ as drawn in Figure 102. In this, the lower end of the object, $A$, and the center of curvature of the surface, $C$, determine the axis of the lens. The axial positions of $A$ and of its image, $a$, are identical with those of $S$ and $S'$ of Figures 98 and 100. These are specified by equation 48. Again, in the diagram we may choose two useful rays to determine the position of the image, $b$, of the object point $B$. The first (1) is a ray parallel to the axis, which upon refraction must pass through the focal point of the

# Image Formation by Refraction from Spherical Surfaces

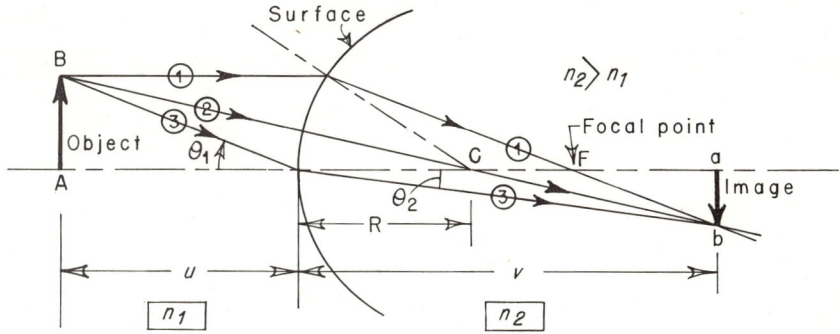

FIGURE 102. The particular rays selected in deriving the formulas for the distances of the object and image from a single refractive surface, and for the magnification.

refractive surface (assuming no spherical aberration). The second (2) is a ray directed to the center of curvature of the surface, for this ray will be undeviated on passing through the surface. The point of intersection of the two rays in the second medium is the position of the image, $b$.

For quantitative discussion it is easier to use a third ray (3), directed to the pole of the surface—that is, to the point of intersection of the axis with the surface. This ray makes an angle of incidence, $\theta_1$, and after refraction an angle $\theta_2$. We may then make use of Snell's law—namely, $n_1 \sin \theta_1 = n_2 \sin \theta_2$. In the diagram we find that

$$\tan \theta_1 = O/u \text{ and } \tan \theta_2 = I/v.$$

If the object $AB$ is not too large for all rays to be considered paraxial, then $\tan \theta_1 \simeq \sin \theta_1$ and $\tan \theta_2 \simeq \sin \theta_2$. By substitution,

$$\frac{I}{O} = \frac{ab}{AB} = -\frac{n_1 v}{n_2 u}, \qquad \ldots (49)$$

which is made minus arbitrarily to indicate that, when all other quantities of the diagram are plus, the image is inverted. This relationship differs from that for the magnification of thin lenses only in the inclusion of the indices of refraction.

The two general formulas for the optical imagery due to refraction from a spherical surface are then

and

$$\left. \begin{array}{l} \dfrac{n_2}{v} + \dfrac{n_1}{u} = \dfrac{n_2 - n_1}{R} = D, \\[6pt] M = -\left[\dfrac{n_1}{n_2}\right]\left[\dfrac{v}{u}\right]. \end{array} \right\} \qquad \ldots (50)$$

These formulas can be used for calculating the position of images in successive centered surfaces, just as for a succession of thin lenses. The image of the first surface acts as the object for the second, and so forth.

## 3. COMBINATION OF TWO REFRACTIVE SURFACES

We can proceed now to the discussion of the properties of an optical unit consisting of two refractive surfaces. An ophthalmic lens is a special case of such a combination of refractive surfaces.

Consider the two refractive surfaces having radii of curvature $R_1$ and $R_2$ (Fig. 103), with indices of refraction of the three successive media being $n_1$, $n_2$, and $n_3$. Of all those rays leaving the object point, $B$, and incident on the first surface, certain ones are selected as construction lines to locate the image formed by the two surfaces. The figure appears rather confusing; so in order to simplify visualization of the course of the selected rays through the combination, it has been redrawn as two figures for the first and second surfaces, respectively (Fig. 104). The image formed by the first surface is taken as the object for the second surface. With this in mind it becomes easy to apply the previously used method of diagramming rays. For the first surface, the lines corresponding to rays 1 and 2 from the object point $B$ are identical in Figures 102 and 104A. These rays of course are subject to refraction by the second surface and no longer lead directly to the final image formed by the second surface. However, the image $a_1b_1$ formed by the first surface (as in Fig. 104A) serves as the object for the second surface. Again we select, for use as construction lines, two particular rays (3 and 4) of all those originating at the point $b_1$ and passing through the second surface (Fig. 104B) to the final image, $b$. Line 3 is that ray directed to $b_1$ parallel to the axis of the lens. This ray, on being refracted by the second surface, proceeds to the focal point, $F_2$, of that surface. Line 4 corresponds to that ray which passes through the center of curvature, $C_2$, of the second surface. This ray will be undeviated on passing through the surface. The point of intersection of these rays, after refraction, is the position of the final image, $ab$.

Thus, we could use the familiar equations to calculate image-

## Image Formation by Refraction from Spherical Surfaces    135

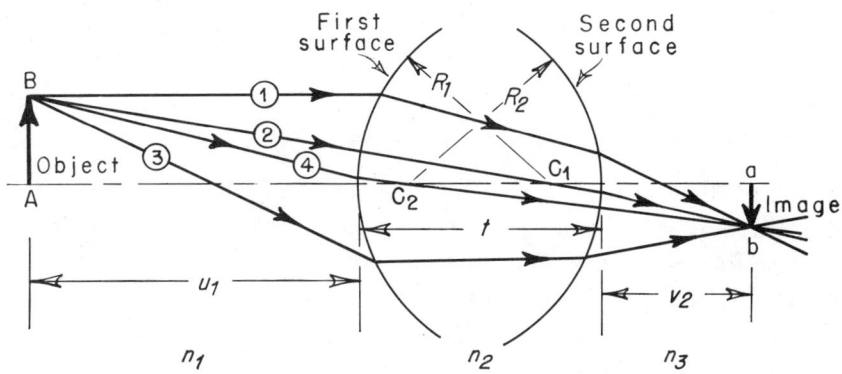

FIGURE 103. Diagram of selected rays from an object to locate the position of the image in a combination of two refractive surfaces.

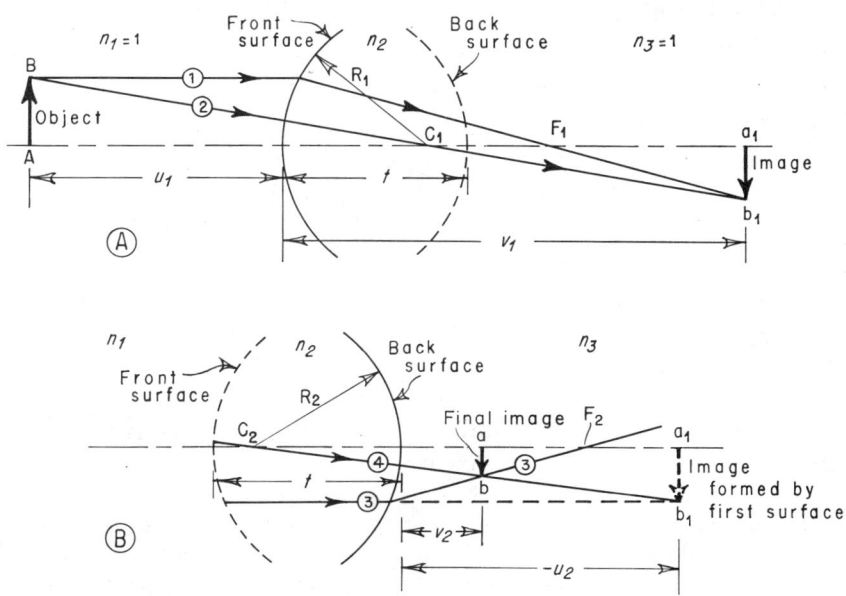

FIGURE 104. Isolated portions of the diagram in Figure 103, illustrating refraction of the rays (A) by the first surface and (B) by the second surface.

object distances successively for the two surfaces, just as for combinations of thin lenses,* namely,

---

*See equations 31, 32, and 33 for the combination of thin lenses in which all quantities are diagrammed as plus.

and
$$\left.\begin{aligned}\frac{n_2}{v_1}+\frac{n_1}{u_1} &= D_1 = \frac{n_2-n_1}{R_1}, \\ u_2 &= t-v_1, \\ \frac{n_3}{v_2}+\frac{n_2}{u_2} &= D_2 = \frac{n_3-n_2}{R_2},\end{aligned}\right\} \quad \ldots (51)$$

to locate the position of the final image. The standard procedure for "tracing" a paraxial ray through a system actually makes use of these formulas. The magnification relationship between the size of the final image and that of the object is

$$M = \frac{I}{O} = \left[\frac{v_1}{n_2}\right]\left[\frac{n_1}{u_1}\right]\left[\frac{v_2}{n_3}\right]\left[\frac{n_2}{u_2}\right].$$

We can follow the same algebraic procedures used for the combination of two thin lenses to arrive at useful formulas for relating the object and image distances, $u$ and $v$, measured from the first and second principal points. We have then

and
$$\left.\begin{aligned}\frac{n_3}{v}+\frac{n_1}{u} &= F, \\ F &= D_1 + D_2 - D_1 D_2 \frac{t}{n_2}, \\ M = \frac{I}{O} &= -\left[\frac{v}{n_3}\right]\left[\frac{n_1}{u}\right], \\ a &= n_1\left[\frac{D_2}{F}\right]\left[\frac{t}{n_2}\right], \\ b &= -n_3\left[\frac{D_1}{F}\right]\left[\frac{t}{n_2}\right].\end{aligned}\right\} \quad \ldots (52)$$

An example of how relationships 51 and 52 may be applied to solve a basic problem is as follows:

Where is the image of the iris (entrance pupil) of the eye and what is its magnification (Fig. 105)? For purposes of illustration this problem will be solved in two ways: (1) by use of the surface-to-surface paraxial-ray tracing (equations 51), and (2) by use of the method of principal points (equations 52). Given are the following: the power of the posterior surface of the cornea ($D_1 = -5.9$ diopters); the power of the anterior surface of the cornea ($D_2 = 48.8$ diopters); the thickness of the cornea ($t = 0.5$ mm $= 0.0005$ meter); the distance between iris and the posterior surface of the cornea ($u_1 = 3.1$ mm $= 0.0031$ meter); the index of refraction of the aqueous humor ($n_1 = 1.336$); the index of refraction of the corneal substance ($n_2 = 1.376$); and the index of refraction of air ($n_3 = 1.000$).

1. By *paraxial-ray tracing* with the use of equations 51. Through the posterior corneal surface

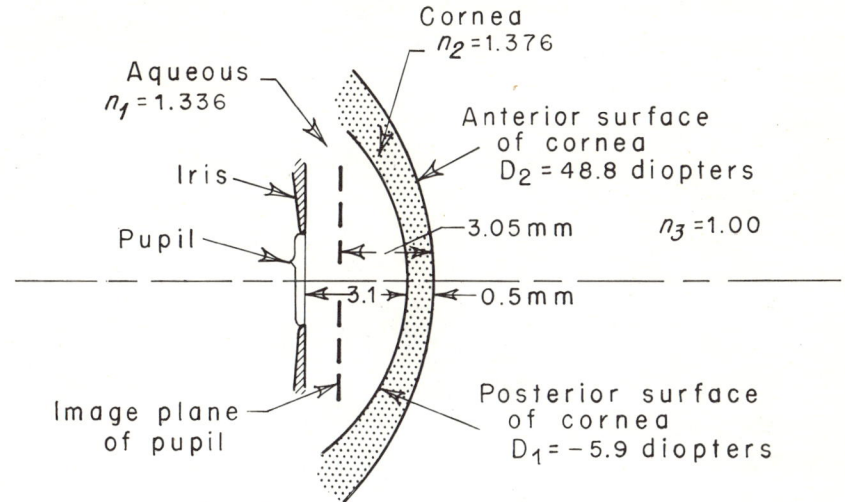

FIGURE 105. Optical system used in finding the position of the entrance pupil (image of real pupil) formed by the cornea of the eye.

$$\frac{1.376}{v_1} + \frac{1.336}{0.0031} = -5.9, \quad \text{or}$$

$$\frac{1.376}{v_1} + 430.9 = -5.9,$$

whence $v_1 = -3.15$ mm.
Through the anterior corneal surface
$$u_2 = 0.5 - (-3.15) = +3.65 \text{ mm.}$$
Then

$$\frac{1}{v_2} + \frac{1.376}{0.00365} = 48.8, \quad \text{or}$$

$$\frac{1}{v_2} + 377.0 = 48.8,$$

whence $v_2 = -3.05$ mm from the anterior surface of the cornea.

The magnification $M$ is
$$M = \left[\frac{-3.15}{1.376}\right]\left[\frac{1.336}{3.10}\right]\left[\frac{-3.05}{1.000}\right]\left[\frac{1.376}{3.65}\right] = 1.135.$$

Thus the pupil appears to be slightly larger than it actually is.

2. By the *principal plane method* with the use of equations 52. The true power of the cornea is

$$F = 48.8 - 5.9 - (-5.9)(48.8)(0.0005)/1.376, \text{ or}$$
$$F = 43.0 \text{ diopters.}$$

The distance of the first principal plane (as drawn in Fig. 105) from the posterior surface of the cornea is given by

$$\frac{a}{1.336} = \left[\frac{48.8}{43.0}\right]\left[\frac{0.0005}{1.376}\right], \text{ whence } a = 0.55 \text{ mm.}$$

The distance of the second principal plane from the anterior surface of the cornea is given by

$$\frac{b}{1.000} = -\left[\frac{-5.9}{43.0}\right]\left[\frac{0.0005}{1.376}\right], \text{ whence } b = +0.05 \text{ mm.}$$

Then the object distance, $u$, $= 3.10 + 0.55 = 3.65$ mm. We have, then,

$$\frac{1}{v} + \frac{1.336}{0.00365} = 43.0,$$

whence the image distance, $v$, $= -3.10$ mm from the posterior principal plane, or $v_b = -3.10 + 0.05 = -3.05$ mm from the anterior surface of the cornea.

Although the actual distance of the iris (pupil) from the cornea is 3.6 mm, it appears to one looking into the eye as only 3.05 mm, about 0.5 mm nearer to the cornea than it actually is. This apparent pupil (the image of the real pupil) seen through the cornea is called the *entrance pupil* of the eye (further discussion of it will be presented later).

The magnification, or the ratio of the size of the image of the pupil to that of the real pupil, is

$$M = -\left[\frac{v}{n_3}\right]\left[\frac{n_1}{u}\right] = -\left[\frac{-3.10}{1.00}\right]\left[\frac{1.3365}{3.65}\right] = 1.135.$$

Thus the image is virtual and erect and is 13.5 per cent larger than the real pupil.

## 4. THICK LENSES

The preceding theory can be applied immediately to a lens in air, for $n_1 = n_3 = 1.00$ (air), and $n_2 = n$ (the index of refraction of glass). Then

$$\frac{1}{v} + \frac{1}{u} = F,$$
$$F = D_1 + D_2 - D_1 D_2 \frac{t}{n}, \quad \quad \ldots (53)$$

and

$$M = -\frac{v}{u}.$$

Now, if the thickness of the lens were negligible ($t = 0$), then the power $F$ of the lens would be $F = D_1 + D_2$. It is essentially this formula that applies when we use a lens (clock) gauge to measure the powers of the two surfaces of a lens. We add the two measurements algebraically to find the power of the lens. By neglecting the thickness we are treating the lens as though it were thin, but for higher powers or high base-curves significant errors may be introduced, so we must consider the lens as a thick lens and include the effect of the thickness.

For ophthalmic lenses, whether we need to take into account the thickness depends upon (1) what precision is required, and (2) whether the magnification properties must be known accurately. Part of the problem also concerns the question of how the powers of ophthalmic lenses shall be specified, especially when precision is necessary and when there must be some degree of standardization. The specification of power depends both on the curvatures of the front and back surfaces and on the thickness. This follows from equations 53, for the effect of including thickness lies in the term $D_1 D_2 t/n$.

How one may use these relationships in actual cases will be illustrated by the following example.

Given a biconvex lens in air, the radius of the front surface, $R_1$, = 6.25 cm; the radius of the back surface, $R_2$, = $-8.33$ cm; thickness = 12.0 mm; and the index of refraction, $n$, = 1.5. There is an object 50 cm from the front surface of the lens. What are the true power of the lens, the positions of the first and second principal points, the position of the image, and the magnification?

The power of the first surface is

$$D_1 = \frac{n_2 - n_1}{R_1} = \frac{1.5 - 1.0}{0.0625} = +8.00 \text{ diopters},$$

and that of the second

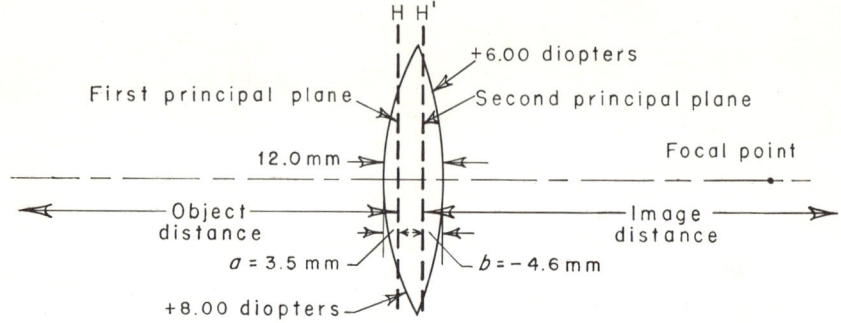

FIGURE 106. Locations of the principal points of a particular, thick lens.

$$D_2 = \frac{n_3 - n_2}{R_2} = \frac{1.0 - 1.5}{-0.0833} = +6.00 \text{ diopters}.$$

The true power of the lens then is

$$F = D_1 + D_2 - D_1 D_2 t/n = 8.00 + 6.00 - (8.00)(6.00)\left(\frac{.0120}{1.50}\right)$$

or

$$F = +14.00 - 0.38 = +13.62 \text{ diopters}.$$

The true focal length, $f$, is the reciprocal of $F$—namely, 7.34 cm.
The distance of the first principal point from the first or front surface is

$$a = (6.00)(12.0)/(13.62)(1.50) = +3.52 \text{ mm}.$$

The distance of the second principal point from the second or back surface is

$$b = -(8.00)(12.0)/(13.62)(1.50) = -4.69 \text{ mm}$$

to the left of the second surface. The positions of these principal points are shown diagrammatically in Figure 106.

The object distance, then, is 50.35 cm, whence (substituting this in equations 53) the image distance is 8.59 cm (which from the back surface is 8.12 cm); the magnification, then, is $-0.171$.

All the previous discussions of optical systems based upon the idea of thin lenses are applicable for thick lenses if we keep in mind the provision that all image and object distances are to be measured from principal points.

The positions of principal points, and therefore the positions of the focal points with respect to the lens itself, vary with the relative curvatures of the surfaces and the thickness of the lens. In many instances depending on the particular shape and thickness of the lens, the order of the principal points may be reversed, in that the second will be in front of the first. Figure 107 illustrates schematically the positions of the principal points for lenses of

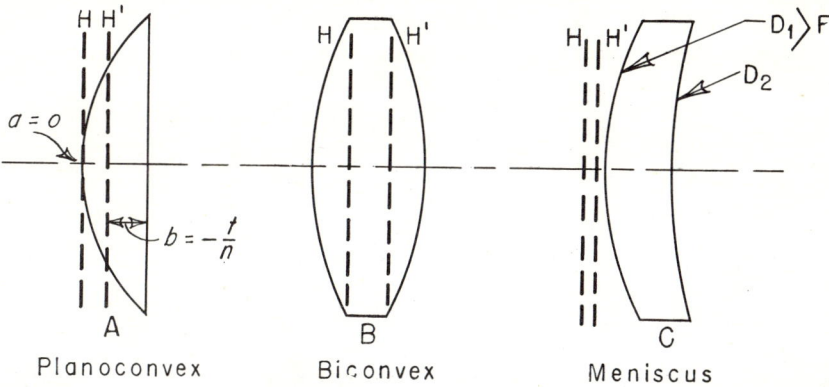

FIGURE 107. Different locations of the principal planes of lenses having the same given power but different shapes.

three different shapes. It is possible for a series of lenses with different shapes to have the same true power.

## EXERCISES

1. Calculate the powers of the glass surfaces in air whose radii of curvature are 40 cm, 250 mm, 10 cm, 70 mm, and 2 cm, if the index of refraction of the glass is 1.50.
2. A glass spherical surface which has a radius of curvature of 20 mm is immersed in water. What is the power of the surface in water? Let the index of refraction of water be 1.333 and that of the glass be 1.50.
3. A planoconvex lens in air has a power of 5 diopters. What is the power of the lens in water?
4. The anterior surface of the cornea of an emmetropic eye has a dioptric power of 48.8. How farsighted would the eye become when in water? The index of refraction of the corneal substance is 1.376; that of water, 1.333.
5. An object 5 cm high is 40 cm from a spherical glass ball with a radius of curvature of 4 cm. How far from the sphere is the image and what is its size? The index of refraction of glass may be taken as 1.50.
6. Calculate the true power of a thick biconvex lens, the radii of curvatures of the front and back surfaces being 5 cm and 20 cm, respectively, the thickness being 12 mm. Where are the principal points?
7. The surfaces of a planoconvex condensing lens have radii of curvatures of infinity and 10 cm, respectively; the lens has a thickness of 3 cm. With its plano side toward the light, so that the rays emerging from the lens will be parallel, how far must this lens be placed from a point source of light?

*Chapter IX*

## OPHTHALMIC OPTICS

### 1. VERTEX POWER OF OPHTHALMIC LENSES

It has been pointed out that in order for a lens to correct fully a myopia or a hyperopia the first focal point of that correcting lens must coincide with the point conjugate to the retina. Clinically, the position of this conjugate point is never determined, except indirectly in terms of the power of the lens needed to correct the ametropia. However, the focal length of the lens (being the reciprocal of the power of that lens) would specify the position of that point.

If the powers of all ophthalmic lenses were specified by the *true* power, $F$, as it has been defined above, then the position of the lens in front of the eye would have to be varied as the shape or thickness of the lens was varied. This follows because the true power is the reciprocal of the true focal length measured from the second principal point, but the position of the second principal point with respect to the lens itself varies with the shape and thickness of the lens. Thus, for their focal points to coincide with the point conjugate to the retina of an ametropic eye, even lenses of the same power (say, 2.00 diopters true power), if their shapes or thicknesses differ, must be at different distances from the eye. This situation is illustrated in Figure 108 for the hyperopic eye and in Figure 109 for the myopic eye.

In the case of lenses of low dioptric power these variations may be relatively unimportant. Nevertheless, for precise ophthalmic optics it is desirable that the lens power be so specified that it will not depend upon the shape of the lens.

It has become customary, accordingly, to specify the dioptric power of ophthalmic lenses as the reciprocal of the *back* focal length—that is, the reciprocal of the distance from the second or

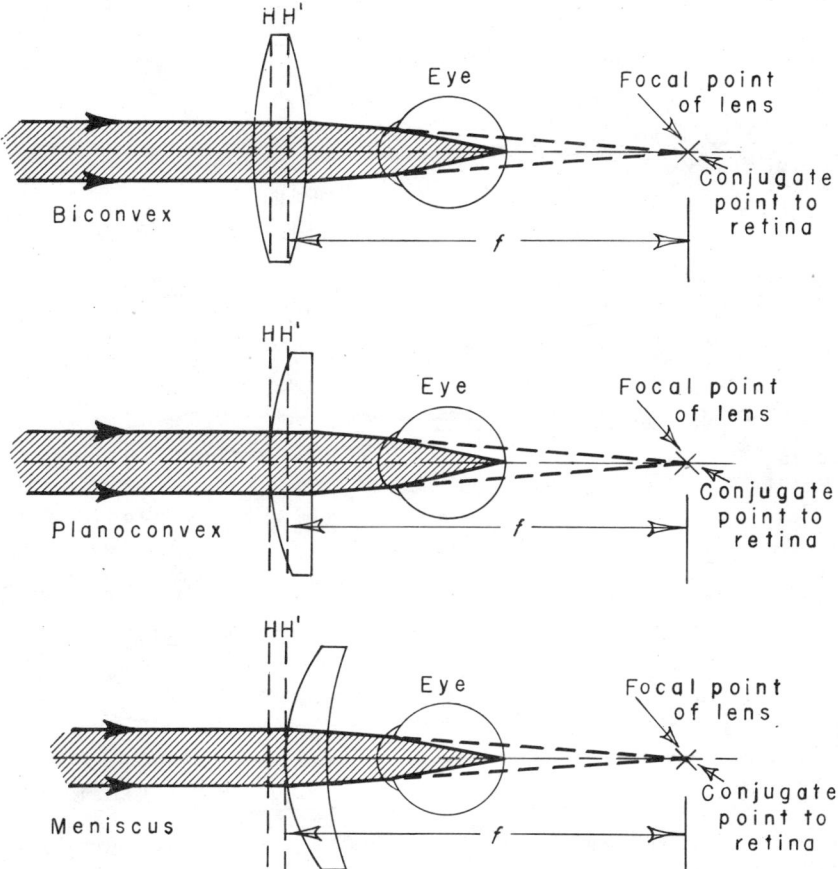

FIGURE 108. Depending upon its shape, a plus lens of a given true power would have to be placed at different distances from the eye in order to correct accurately a hyperopic refractive error.

back surface of the lens to the first focal point—a distance which can be measured easily. This power is called the *vertex power* of the lens because it refers to a power measured from the *vertex* of the lens. Figure 110 illustrates the difference between the true and back focal lengths with respect to a meniscus type of converging lens. The powers of all modern ophthalmic lenses, both trial-case and corrective, are now specified and measured in terms of vertex power.

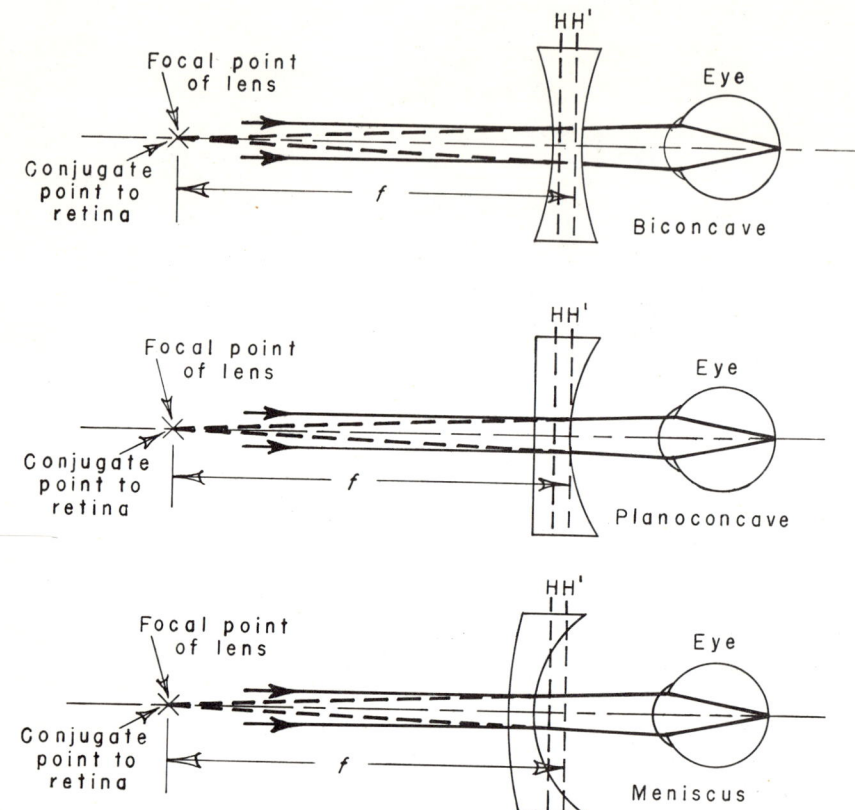

FIGURE 109. Depending upon its shape, a minus lens of a given true power would have to be placed at different distances from the eye in order to correct accurately a myopic refractive error.

The immediate problem in this discussion is to find the relationships between the vertex power and the powers of the first and second surfaces and the thickness of the lens. Referring to the basic formulas for tracing a paraxial ray through a lens (equations 51), we find that when the object distance, $u_1$, is very large (infinite), the final image distance, $v_2$, is equal to the back focal length of the lens, $f_v$. For an ophthalmic lens $n_1 = n_3 = 1.00$ (air), and $n_2 = n$, index of refraction of the glass. Thus, taking the vergence of the incident light $n_1/u_1 = 0$ (zero), the three equations to be used are

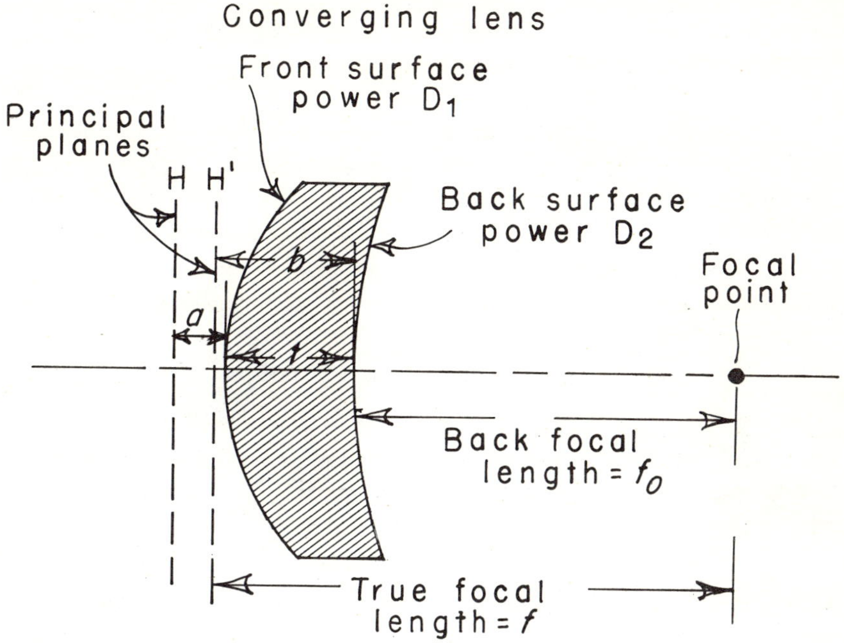

## Converging lens

FIGURE 110. Relationship between true focal length and back focal length of an ophthalmic lens.

$$\frac{n}{v_1} = D_1,$$
$$u_2 = t - v_1, \text{ and} \qquad \qquad \ldots (54)$$
$$\frac{1}{f_o} + \frac{n}{u_2} = D_2,$$

in which, as before, $D_1$ and $D_2$ are the surface powers of the front and back surfaces, respectively, and $t$ is the thickness. The distance $v_2 = f_o$ is the back focal distance. When $u_2$ is eliminated from the above three equations, the *vertex power*, $V_o = 1/f_o$ diopters, is given by

$$V_o = \frac{D_1}{1 - D_1 t/n} + D_2, \text{ diopters.} \qquad \ldots (55)$$

The thickness, $t$, must be expressed in meters. This formula shows that the effective power of the front surface, as affected by the thickness of the lens, is $D_1/(1 - D_1 t/n)$. The fraction $1/(1 - D_1 t/n)$ is called a *shape factor* of the lens because it depends only upon the power of the front surface and the thickness of the lens.

Although the vertex power is an advantageous designation to use in specifying the power of ophthalmic lenses, it should be clear that it pertains only to a distant object and cannot be used as the power of the lens in those formulas which state the relationships between object and image distances or in the formula specifying the magnification of the image. These distances would have to be measured from the principal points and the true power of the lens used. However, vertex power and true power are related, for if equation 55 is written with a common denominator—namely,

$$V_o = \frac{D_1 + D_2 - D_1 D_2 t/n}{1 - D_1 t/n} = \frac{F}{1 - D_1 t/n}, \quad \ldots (56)$$

we note that the numerator is identical to the formula for the true power of the lens (equation 53). Thus the true power, $F$, is related to the vertex power, $V_o$, by

$$F = V_o(1 - D_1 t/n).$$

Again, we note that if the thickness of the lens can be neglected, then $F = V_o = D_1 + D_2$; that is, the two powers are the same, and the lens is effectively a thin lens.

For practical purposes, the expression for vertex power (equation 55) can be simplified. The shape-factor fraction, $1/(1 - D_1 t/n)$, can be divided out longhand into a power series. Thus,

$$\frac{1}{(1 - D_1 t/n)} = 1 + D_1 t/n + (D_1 t/n)^2 + (D_1 t/n)^3 + \ldots \quad \ldots (57)$$

The expression for the vertex power, after substitution in equation 55, is

$$V_o = D_1[1 + (D_1 t/n) + (D_1 t/n)^2 + \ldots] + D_2,$$

which can be written

$$V_o = D_1 + D_2 + D_1(D_1 t/n) + D_1(D_1 t/n)^2 + \ldots.$$

Now, the quantity $D_1(t/n)$ is never very large, so all terms beyond the third can be neglected. To illustrate this, suppose the power of the front surface of a lens is +10.00 diopters and the thickness is 6 mm. Then the third term is

$$[10]\ [10]\ [.006/1.5] = 0.40 \text{ diopter},$$

and the fourth term is

$$[10]\ [(10)\ (.006/1.5)]^2 = 0.016 \text{ diopter},$$

a power error which in ophthalmic optics is certainly negligible. The formula for the vertex power of a lens becomes simply

$$V_o = D_1 + D_2 + D_1^2 t/n. \qquad \ldots (58)$$

This shows that when the thickness is taken into account, the lens has more *plus* vertex power than that given by the sums of the powers of the two surfaces alone. The last term, $[D_1^2 t/n]$, is sometimes called the *allowance factor* and depends only upon the dioptric power of the front surface and the thickness of the lens. Optical firms usually supply tables which give this allowance factor in one form or another. The following table is an excerpt from a more extensive table giving the allowance factor in diopters for lenses which have given front surface powers and given thicknesses.

| Thickness, mm | Power of front surface —diopters | | | | | |
|---|---|---|---|---|---|---|
| | 4.00 | 6.00 | 8.00 | 10.00 | 12.00 | 14.00 |
| 2 | .02 | .05 | .08 | .13 | .21 | .26 |
| 3 | .03 | .07 | .13 | .20 | .29 | .40 |
| 4 | .04 | .10 | .17 | .27 | .39 | .53 |
| 5 | .05 | .12 | .22 | .34 | .49 | .67 |
| 6 | .06 | .14 | .26 | .40 | .59 | .82 |

This table is useful in precise ophthalmic optics. The vertex power of a lens can be obtained by measuring the powers of the front and back surfaces with a clock surface gauge and adding these two powers algebraically (equation 58), then, from the thickness measured by a caliper, finding the corresponding allowance factor from the table and adding that to the sum. More practical use of such a table is made by the manufacturing optician, for the vertex power is specified by the prescription, and the surface powers must be determined. He decides on the front surface power and thickness and then calculates (or reads from prepared tables) what the power of the back surface must be—namely, $D_2 = V_o - D_1 - (D_1^2 t/n)$. The back surface must always have a slightly more minus power, the difference being the allowance factor.

The lensometer (vertometer and so on) is so designed as to measure directly the vertex power of ophthalmic lenses (see Chapter XIII, section 4).

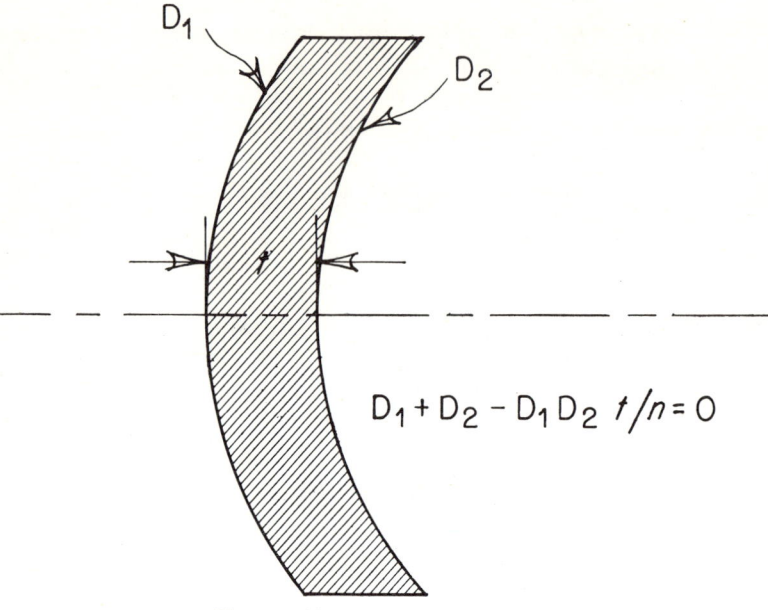

FIGURE 111. The afocal lens.

## 2. AFOCAL LENSES

A special category of ophthalmic lenses includes those for which the powers are zero.[6] These lenses are telescopic lenses; parallel rays of light incident upon them also emerge parallel. Such lenses are of necessity Galilean in type, in that the minus power of the back surface is such as to offset the power and shape factor of the front surface (Fig. 111). However, these lenses do produce an angular magnification of objects viewed through them. Such lenses have been used in aniseikonic prescriptions to introduce magnification and thus to equalize the sizes of the images in the two eyes. These lenses are sometimes called *iseikonic*. On the basis of the discussion of telescopic systems with thin lenses (Chapter VII, section 3), we can write immediately

$$D_1 + D_2 - D_1 D_2 t/n = 0, \qquad \ldots (59)$$

and the angular magnification

$$A = -\frac{D_2}{D_1} = 1/(1 - D_1 t/n). \qquad \ldots (60)$$

Since the power of the lens (both true power and vertex power) is zero, the principal planes lie at infinity. The angular magnification is equal to the shape factor of the lens.

The front surface powers of lenses in general can rarely be made greater than 18 diopters, and the thicknesses can rarely be greater than 8 mm; thus the amount of angular magnification that can be obtained with single afocal lenses is limited. Because the quantity $D_1 t/n$ is small compared to unity, the angular magnification can be written

$$A = 1 + D_1 t/n, \qquad \ldots (61)$$

in which higher powers of $(D_1 t/n)$ are neglected, as was shown in equation 57. Thus the maximal practical angular magnification would be $A = 1 + (18)(.008/1.5)$, or $A = 1.09$, or 9 per cent. Actually, even this lens would be heavy and generally unsuitable for use. Usually we cannot expect to find a practical afocal lens with an angular magnification greater than 5 per cent. Therefore such lenses are not used to equalize the images in the two eyes in unilateral aphakia when corrected with spectacle lenses, for the difference in the magnification of the images in the two eyes may be as much as 30 per cent.

## 3. NODAL POINTS

The glass lens is immersed in air; hence the index of refraction ($n_1$) on the object side of the lens is the same as that ($n_3$) on the image side. Suppose, however, that the index of refraction on the image side of the lens, $n_3$, is different from that, $n_1$, on the object side. Then the relationship between the positions of the object and of the image is a little more complicated. This is the situation for the eye, where the image space is in the vitreous humor: The anterior and posterior focal lengths are now *not* the same.

For the lens with air on both sides, a ray directed to the first principal point will emerge from the system as if from the second principal point, both the incident and the emergent rays making the same angle to the optic axis of the lens. When the indices of refraction on the image and object sides of the refractive surfaces are different, this statement no longer holds.

There are, fortunately, two other special points on the axis of the system which serve the same purpose; that is to say, in such

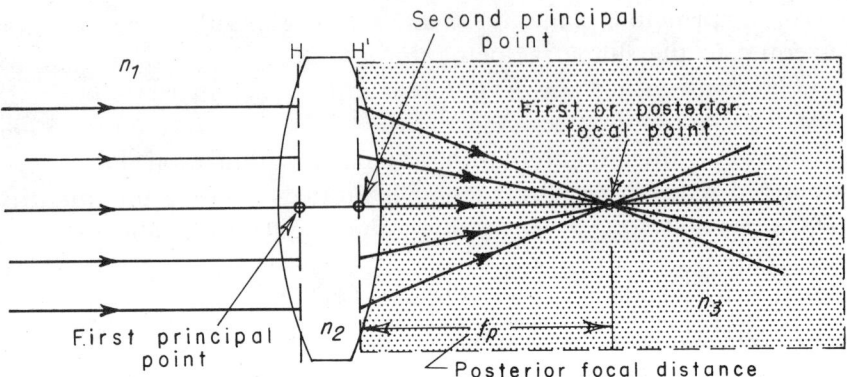

FIGURE 112. The anterior and posterior focal lengths of a lens are different if the index of refraction of the image space differs from that of the object space. This figure illustrates the rays which determine the posterior focal distance.

an optical system a ray from the object which is directed to the first of these points will emerge from the system as if from the second point, and both rays will make the same angle of inclination to the optic axis. These particular reference points are called *nodal* points; they occur only in optical systems, such as that of the eye, in which the index of refraction on the image side differs from that on the object side. With ordinary lenses in air the nodal points are identical to the principal points.

It has been shown that in any centered optical system the object and image distances, $u$ and $v$, measured from the first and second principal points, respectively, are related by equation 52—namely,

$$\frac{n_3}{v} + \frac{n_1}{u} = F,$$

where, as before, the true power

$$F = D_1 + D_2 - D_1 D_2 t / n_2.$$

The distance of the first focal point from the second principal point, called the *posterior* focal length, can be found algebraically from equation 52 simply by letting the distance, $u$, to the object be very large (see Fig. 112)—that is, making $u \to \infty$. Then the corresponding image distance, $v$, will be identical with the true posterior focal length, $f_p$, of the system—namely,

$$v \Big]_{u \to \infty} = f_p = \frac{n_3}{F} = n_3 f. \qquad \ldots (62)$$

# Ophthalmic Optics

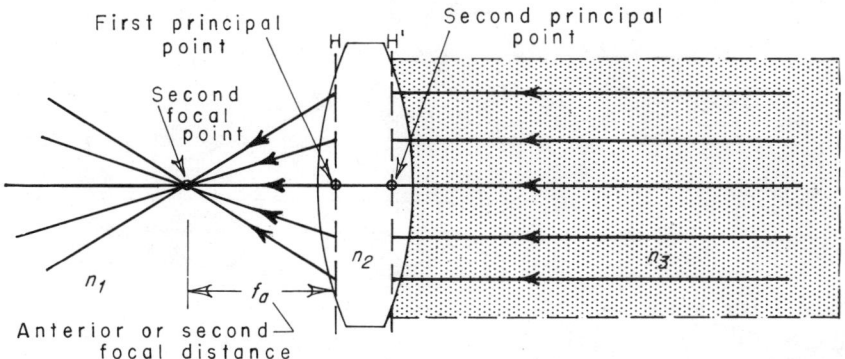

FIGURE 113. The anterior and posterior focal lengths of a lens are different if the index of refraction of the image space differs from that of the object space. This figure illustrates the rays which determine the anterior focal distance.

This states that the true posterior focal length is the reciprocal of the true power multiplied by the index of refraction of the image space.

Similarly, the true *anterior* focal length, $f_a$, can be found by setting the image distance $v \to \infty$ (infinity). Then (Fig. 113)

$$u\Big]_{v\to\infty} = f_a = \frac{n_1}{F} = n_1 f. \qquad \ldots (63)$$

This states that the true anterior focal length is the reciprocal of the true power multiplied by the index of refraction of the object space.

Now the relationship between the anterior and posterior focal lengths is obtained by eliminating $F$ from equations 62 and 63. Thus we obtain

$$f_a = \frac{n_1}{n_3} f_p$$

or

$$\frac{f_a}{n_1} = \frac{f_p}{n_3} = \frac{1}{F}. \qquad \ldots (64)$$

If the index of refraction on the image side, $n_3$, of the system is higher than that on the object side, $n_1$, the true posterior focal distance on the image side is greater than that on the object side. This is true in the case of the eye, in which the index of refraction of the vitreous humor is $n_3 = 1.333$ and that of air is $n_1 = 1$.

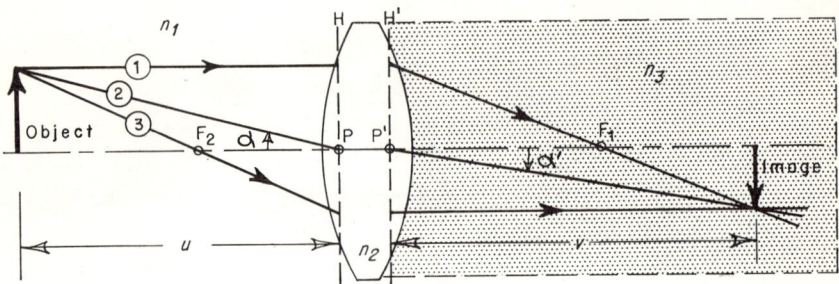

FIGURE 114. Relationship between the angles subtended by the object and the image at the principal points for a lens system in which the index of refraction is greater in the image space than in the object space.

The first or posterior focal distance is about 22.7 mm, whereas the anterior focal distance is about 17 mm, the two distances being measured from the second and first principal points of the eye, respectively.

Figure 114 shows the selected rays from a given object point that are useful for diagramming the position of the image in such an optical system. Rays 1 and 3 determine the position of the image. Ray 2 which is directed to the first *principal* point makes an angle $\alpha$ with the axis and leaves the system as if from the second *principal* point, but making a different angle $\alpha'$ with the axis.

The ratio of the size of the image, $I$, to the size of the object, $O$—that is, the magnification—is equation 52.

$$M = \frac{I}{O} = -\left[\frac{v}{n_3}\right]\left[\frac{n_1}{u}\right] = -\left[\frac{n_1}{n_3}\right]\left[\frac{v}{u}\right]. \quad \ldots (65)$$

Since $\tan \alpha = O/u$ and $\tan \alpha' = I/v$, we have

$$\frac{\tan \alpha'}{\tan \alpha} = \left[\frac{I}{O}\right]\left[\frac{u}{v}\right] = -\frac{n_1}{n_3}. \quad \ldots (66)$$

Thus, with considerable precision, the ratio of the angle subtended by the image at the second principal point to the angle subtended by the object at the first principal point is equal to the ratio of the first and last indices of refraction. If $n_3 > n_1$, as for the eye, in which the index of refraction of the vitreous humor is greater than that of air, then the *angle* subtended at the posterior principal point by the image is smaller than that subtended at the first principal point by the object.

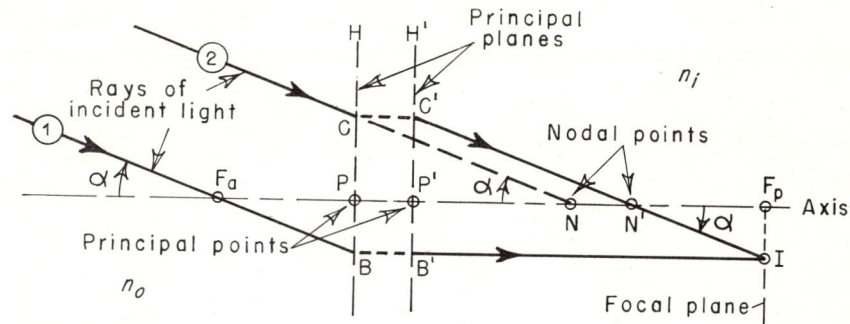

FIGURE 115. Construction of rays selected to locate the positions of the two nodal points. At the nodal points the angle subtended by the object is the same as that subtended by the image.

## 4. LOCATION OF NODAL POINTS

In any optical system, the positions of the two nodal points relative to the positions of the principal planes and of the focal points can be determined diagrammatically by the method shown in Figure 115. By definition and experiment, an incident ray directed toward the first nodal point emerges as the refracted ray from the system as if proceeding from the second nodal point, both rays making the same angle with the axis—that is, $\alpha' = \alpha$. In this figure, suppose $P$ and $P'$ are the first and second principal points and $H$ and $H'$ are the first and second principal planes of an optical system. The anterior and posterior focal points are $F_a$ and $F_p$, respectively; and the posterior focal distance, $P'F_p$, is taken greater than the anterior focal distance, $PF_a$, corresponding to the case in which the index of refraction of the image space, $n_i$, is greater than that, $n_o$, of the object side of the optical system.

Consider a beam of parallel light rays from a very distant object incident to the optical system at an angle $\alpha$ with the axis of the system. Let us select a ray (1) from this bundle which passes through the anterior focal point, $F_a$. This ray then effectively intersects the first principal plane, $H$, at a point, $B$. For diagrammatic purposes, this ray leaves the second principal plane, $H'$, at the same distance below the axis, $B'$, and proceeds parallel to the axis, intersecting the posterior focal plane (the plane at $F_p$) at the image point, $I$. Now, of all the rays incident to the system, there is one which, on passing through the system to the image, $I$, also makes an angle $\alpha$ to the optic axis in the image space. This ray, $C'I$, is drawn parallel to $F_aB$; it intersects the axis at the point $N'$, which is the second nodal point. The incident ray (2) corresponding to $C'I$ is that one which strikes the first principal plane, $H$, at a point, $C$, the same height above the axis as $C'$. The line representing ray 2, when extended, intersects the axis at $N$, the first nodal point. This ray (2) then intersects the axis at the angle $\alpha$, as does the emergent ray, $C'I$. The poinst, $N$ and $N'$, are therefore the first and second nodal points of the system, respectively. A ray directed toward the first nodal point, $N$, emerges from the system as if from the second nodal point, $N'$, the incident and emergent rays making the same angle with the optical axis of the system.

We can determine the positions of the nodal points by the following consideration. In Figure 115 the right triangles $F_aPB$ and $N'F_pI$ are equal because the angles and opposite sides are equal; hence $F_aP = N'F_p$. That is, [*the anterior focal distance is equal to the distance from the second nodal point to the posterior focal point*]. Similarly, the separation of the nodal points, $NN'$, is equal to the separation of the principal points, $PP'$, because triangles $CPN$ and $C'P'N'$ are equal.

We can write that the distance $P'F_p = P'N + NN' + N'F_p$. However, since $NN' = PP'$, and $N'F_p = F_aP$, one has, on substitution, $P'F_p = P'N + PP' + F_aP = F_aN$. That is to say, [*the distance of the anterior focal point from the first nodal point is equal to the posterior focal distance*].

All this seems complicated, but now we have derived from it the rule for locating the nodal points. The only place in ophthalmic optics in which this discussion is pertinent is the dioptrics of the human eye, where the index of refraction of the vitreous humor (image space) is different from that of air in contact with the cornea (object space).

## 5. THE CARDINAL POINTS

Under the general category of *cardinal points* are grouped all six of these important reference points of any optical system:

1. The anterior (second) and posterior (first) *focal* points,

2. The first and second *principal* points.
3. The first and second *nodal* points.

## 6. THE SCHEMATIC EYE

A schematic eye, or a model, whose dimensions and optical properties approximate those of the ordinary living eye is useful both in ophthalmologic problems and in the study of physiologic optics. Such a model enables us to visualize more readily the optical properties of the eye and the location of foreign bodies as well as to determine the size of the retinal image.

Many schematic eyes have been devised in the history of physiologic optics (Listing,[4] Donders,[1] Tscherning,[8] von Helmholtz,[3] Gullstrand,[2] and others), each of which is different in some particular, although basically all have much in common. The more recent statistical studies on the dimensions of human eyes (Stenström,[7] among others) should perhaps furnish a new basis for the selection of the dimensions for a schematic eye.

One of the difficulties, from the optical point of view, is that certain approximations are necessary; hence the model must be made an idealized version of the actual eye. The optical properties of the average living eye differ from those of a schematic eye in the following ways:

1. The refractive surfaces tend to be slightly aspheric. Even in nonastigmatic eyes only the central areas of the surfaces can be said to be spherical. The corneal surface tends to flatten toward the limbus.
2. The crystalline lens is usually slightly decentered and tipped with respect to the axis of the cornea and with respect to the visual axis of the eye. Thus, the actual eye is not a centered optical system.
3. Of importance also is the fact that the crystalline lens consists of a nonhomogeneous material, which makes simulation of this lens in a schematic eye difficult. The structure of the lens is laminar, and the index of refraction increases toward the center of the lens. Matthiessen[5] determined the index of refraction of successive layers in the laminar structure and represented the results of the measurements by a

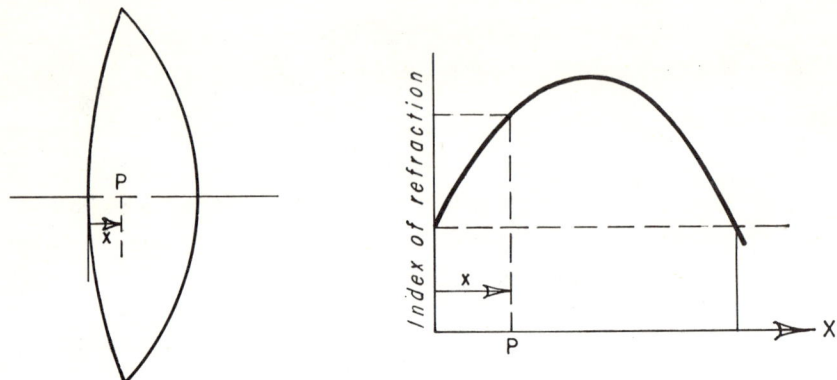

FIGURE 116. Schematic illustration of the progressive change of the index of refraction of the crystalline lens of the eye.

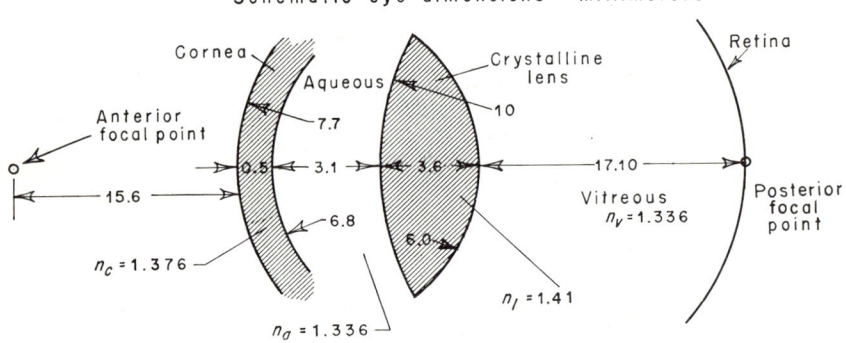

FIGURE 117. Dimensions (mm) of a useful schematic eye based upon that of Gullstrand (not drawn to scale).

parabola (Fig. 116). Gullstrand incorporated a so-called core lens in his schematic eye to meet this problem and also to simulate the nucleus of the living crystalline lens.

For the most part, a schematic eye based upon that of Gullstrand is adequate if it be somewhat modified by elimination of the core and by making the eye emmetropic. The depth of the anterior chamber also is increased. The dimensions of this eye are illustrated in Figure 117. The locations of the cardinal points and the dioptric distances are given in Figure 118. An effective index of refraction of 1.41 is used for the crystalline lens. The optical

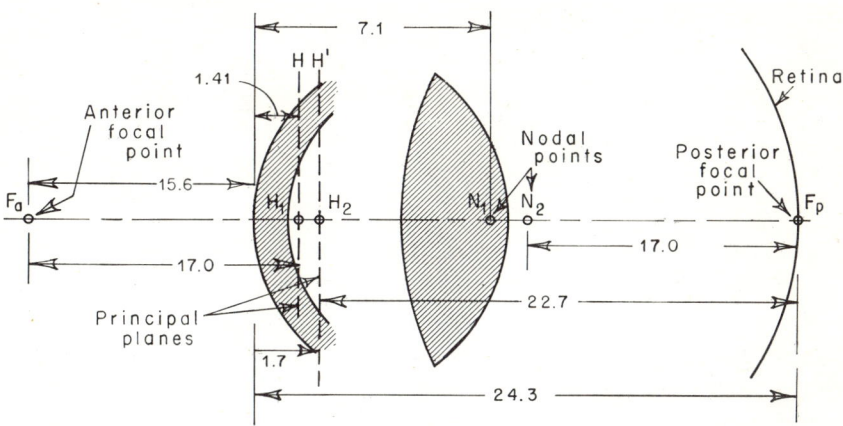

FIGURE 118. The cardinal points and dioptric distances (mm) of the schematic eye (not drawn to scale).

characteristics are given here only for the unaccommodated or relaxed eye.

The dioptric properties of the schematic eye are given in the following table.

### DIOPTRIC CONSTANTS OF THE SCHEMATIC EYE

| Components | | Whole Eye | |
|---|---|---|---|
| Corneal system | | Power of eye | 58.9 diopters |
|   Power | 43.00 diopters | Posterior focal length | 22.7 mm |
|   Power front surface | 48.80 diopters | Anterior focal length | 17.0 mm |
|   Power back surface | −5.90 diopters | | |
|   First principal point | −0.05 mm | *From cornea* | |
|   Second principal point | −0.50 mm |   Posterior focal point | 24.3 mm |
| | |   Anterior focal point | 15.6 mm |
| Lens system | |   First principal point | 1.4 mm |
|   Power | 19.50 diopters |   Second principal point | 1.7 mm |
|   Power front surface | 7.40 diopters |   First nodal point | 7.1 mm |
|   Power back surface | 12.30 diopters |   Second nodal point | 7.4 mm |
|   First principal point | 2.10 mm |   Entrance pupil | 3.0 mm |
|   Second principal point | −1.30 mm |   Exit pupil | 3.7 mm |
| | |   Second nodal point from retina | 17.0 mm |

## 7. SIZE OF RETINAL IMAGE

Knowledge of the position of the nodal points of the eye permits us to calculate the relative sizes of an object and its image on

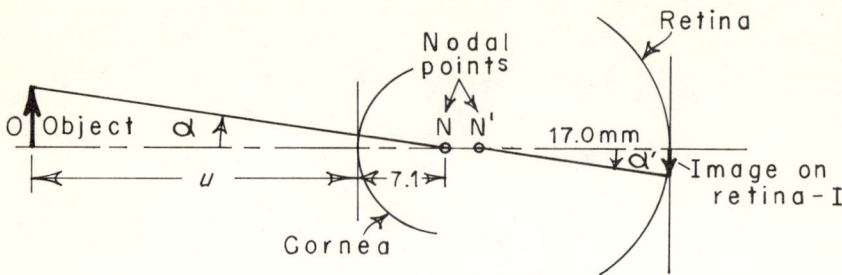

FIGURE 119. The particular dimensions (mm) of the schematic eye that are useful in calculating the size of the retinal image of an object in space.

the retina. In Figure 119, since the angle subtended by an object at the first nodal point is equal to the angle subtended by the image at the second nodal point, the two right triangles $O$-$N$ and $I$-$N'$ are similar; hence the sides of these triangles are proportional. Thus the ratio of the size of the image to that of the object is

$$I/O = 17.0/(u + 7.1), \qquad \ldots (67)$$

in which the object distance from the cornea, $u$, is expressed in millimeters, since the distance from the second nodal point to the retina is also expressed in millimeters (17.0). Usually, object distances are large so that the additive distance, 7.1 mm, can be neglected.

*Example:* On the tangent screen at a distance of 1 meter from the eye, a circular scotoma 10 cm in diameter is mapped out. What is the size of the scotoma on the retina? One has, from equation 67, $I = (10)(17.0)/(1007.1) = 0.17$ cm $= 1.7$ mm.

## 8. THE REDUCED SCHEMATIC EYE

It is sometimes convenient, for certain approximate calculations, to have a schematic eye reduced to an equivalent single refracting surface and a single ocular medium. To obtain an unaccommodated schematic eye with approximately the same refractive power and the same overall dimensions as that more complete schematic eye just discussed, the radius of the corneal surface must be considerably less than that for the normal schematic eye. Figure 120 illustrates the dimensions of a reduced schematic eye. For such an eye, the posterior focal length is $f_p = n_2 R/(n_2 - 1)$; the anterior focal length, measured from the cornea, is $f_a = -R/(n_2 - 1)$. The

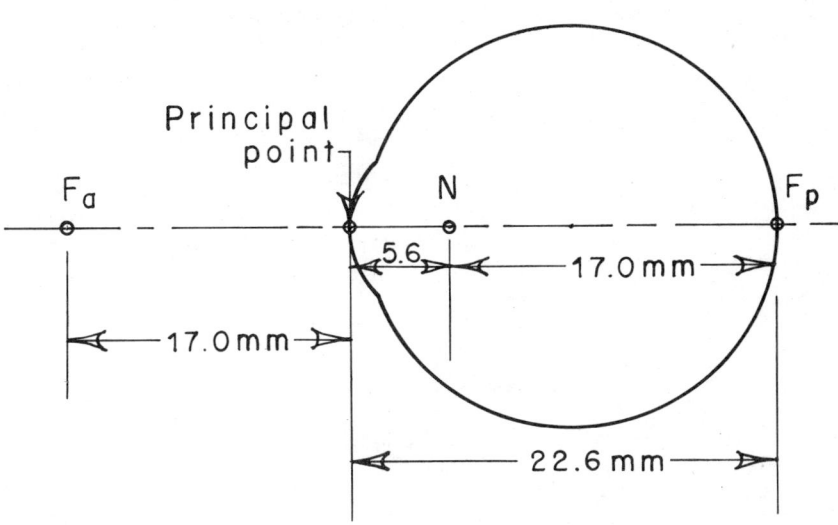

FIGURE 120. Dimensions of a reduced schematic eye.

first and second principal planes are at the cornea, and the nodal points collapse into one, at a distance from the cornea equal to the radius of curvature of the corneal surface. (A ray directed to the center of curvature of a refractive surface is undeviated.) The dioptric constants are given in the following table.

DIOPTRIC CONSTANTS OF THE REDUCED SCHEMATIC EYE

| | |
|---|---|
| Radius of curvature of cornea | 5.65 mm |
| Axial length | 22.60 mm |
| Index of refraction of ocular media | 1.33 (water) |
| Power of eye | 58.90 diopters |
| First principal plane from cornea | 0.00 |
| Nodal point from cornea | 5.65 mm |
| Anterior focal point | 17.00 mm |
| Posterior focal point | 22.60 mm |
| Nodal point from retina | 17.00 mm |

## EXERCISES

1. A trial-case sphere has a front surface power of 6.50 diopters, a thickness of 4 mm, and a back surface power of −12.00 diopters. (Index of refraction is 1.50.) Calculate the vertex power. What is the true power of the lens? Reverse the lens (minus surface in front) and calculate the vertex power.

2. Calculate true and vertex powers for each of the following lenses (take index of refraction = 1.50).

|     | $D_1$  | $D_2$   | $t$(mm) |
|-----|--------|---------|---------|
| (a) | +6.00  | −8.00   | 1.5     |
| (b) | +10.00 | −4.00   | 4.0     |
| (c) | +15.00 | −8.00   | 5.0     |
| (d) | +8.00  | −10.00  | 2.0     |

In the above examples, if the lenses are turned around (interchange $D_1$ and $D_2$), what would be their vertex powers? True powers?

3. Calculate the allowance factor for lenses having the following powers of front surface and thickness:

|     | $D_1$       | $t$   |
|-----|-------------|-------|
| (a) | 6 diopters  | 2 mm  |
| (b) | 10 diopters | 2 mm  |
| (c) | 10 diopters | 4 mm  |
| (d) | 15 diopters | 5 mm  |

4. Calculate the dioptric surface power of the second surface for the following vertex powers and thicknesses of lenses:

|     | $V_o$       | $D_1$           | $t$   |
|-----|-------------|-----------------|-------|
| (a) | −4 diopters | 4.50 diopters   | 1 mm  |
| (b) | +5 diopters | 8.50 diopters   | 3 mm  |
| (c) | +10 diopters| 12.00 diopters  | 5 mm  |

5. Calculate the angular magnification of an afocal lens that has a front surface power of +10.00 diopters and a thickness of 4.5 mm. What should be the power of the second surface? Assume that the index of refraction of the glass is 1.50. What is the magnification in per cent?

6. An emmetropic (but completely presbyopic) eye views an object at 1 meter from the cornea. How far from the retina would the image be focused? That is to say, for each diopter of ametropia (if not too high), how many millimeters does the image lie before or behind the retina?

7. The angular width of the blind spot is 6° (tan 6° = 0.105).
   (a) What is the width of the blind spot on the retina?
   (b) What is the width of area on a tangent screen at 2 meters?
   (c) What is the width of area on a tangent screen at 6 meters?

8. The separation of the headlights on a car is about 150 cm. When the car is 30 meters (90 feet) away, what is the separation of the images of the lights on the retina?

9. The resolving power of the eye is 1 minute of arc (tan 1' = 0.00029). What is the corresponding dimension on the retina?

# REFERENCES

1. DONDERS, F. C.: *On the Anomalies of Accommodation and Refraction of the Eye,* trans. by W. D. Moore. London, New Sydenham Society, 1864, p. 173.
2. GULLSTRAND, A.: In von Helmholtz, H.: *Helmholtz's Treatise on Physiological Optics,* trans. from the 3rd German ed., J. P. C. Southall (Ed.). Lancaster,

The Optical Society of America, 1924, vol. 1, p. 350. Reprinted 1964, New York, Dover.
3. von Helmholtz, H.: *Helmholtz's Treatise on Physiological Optics*, trans. from the 3rd German ed., J. P. C. Southall (Ed.). Lancaster, The Optical Society of America, 1924, vol. 1, p. 152. Reprinted 1964, New York, Dover.
4. Listing, J. B., quoted by von Helmholtz, H.: *Ibid.*, p. 95.
5. Matthiessen, L., quoted by von Helmholtz, H.: *Ibid.*, pp. 340-344.
6. Ogle, K. N.: *Researches in Binocular Vision.* Philadelphia, Saunders, 1950, pp. 121-132. Reprinted 1964, New York, Hafner.
7. Stenström, Sölve: Investigation of the variation and the correlation of the optical elements of human eyes (trans. by Daniel Woolf). *Amer J Optom*, Monograph 58, 1948.
8. Tscherning, M.: *Physiologic Optics: Dioptrics of the Eye, Functions of the Retina, Ocular Movements and Binocular Vision*, trans. by Carl Weiland. Philadelphia. Keystone Books, 1920, pp. 33-46.

*Chapter X*

## ASTIGMATISM

### 1. ASTIGMATIC IMAGERY

If the image of a point object, as formed by an optical system, is itself a point, the imagery is said to be "stigmatic." In most optical systems—and certainly in the case of simple lenses—this stigmatic imagery is not generally obtained because there are usually a number of aberrations as well as diffraction effects which cause some deterioration in the image. The term *astigmatic*, meaning *not stigmatic*, is reserved, however, to describe a particular kind of aberration or deviation from stigmatic imagery—namely, that caused by the fact that the refractive surface (or surfaces) of the lens system does not act as a spherical surface but as one that may be described by *two* curvatures, which by the geometry are always at right angles to each other.

In ophthalmic optics this type of defect in image formation plays a prominent role, primarily because in a majority of human eyes the corneal surfaces are not spherical but are toroidal, in that each acts as if it had *two* different curvatures in meridians at right angles to each other. An eye with this defect is said to exhibit astigmatism because the imagery is astigmatic.* The term *toric* comes from the three-dimensional geometric figure called the torus, which is a ring, any point on the surface of which lies on two circles of different radii (Fig. 121). Examples of toric surfaces, besides that of a torus itself, would be a limited area on the surface of an automobile tire, a doughnut (perhaps) or—as degenerate examples—a football or a lemon, and in the extreme case, a simple geometric cylinder.

Every point on a toric optical surface lies on two circles with

---

*The surface of the cornea may also be slightly warped, in which case the astigmatism is said to be *irregular*. If the central portion of the surface is indeed practically toroidal, then the astigmatism is said to be *regular*. In this discussion we are interested only in the problem of regular astigmatism.

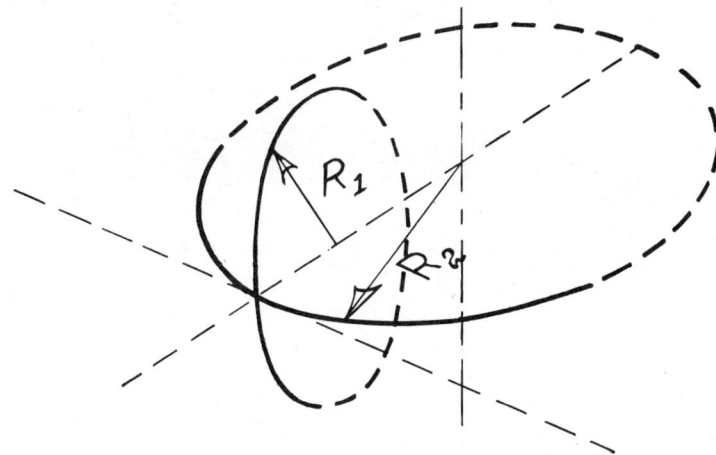

FIGURE 121. Geometric figure illustrating the two radii of curvature of a torus.

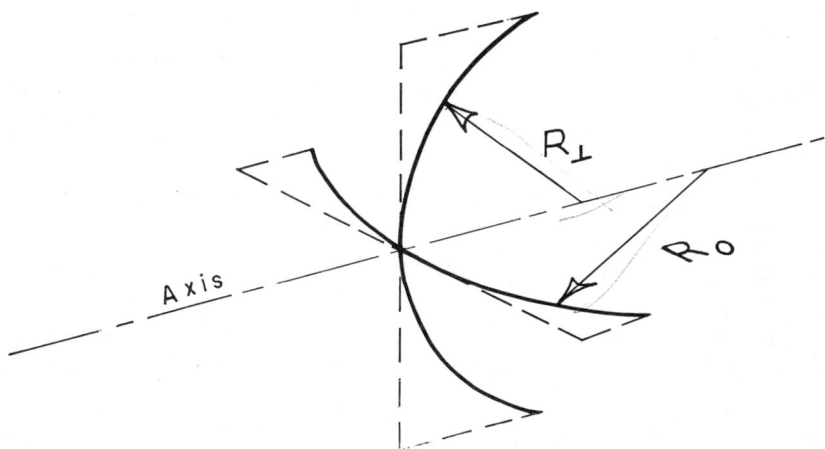

FIGURE 122. Geometric figure illustrating the two radii of curvature at a point on a toric surface.

radii of curvature $R_o$ and $R_\perp$; hence there are two dioptric powers—namely, $D_o = (n_2 - n_1)/R_o$ and $D_\perp = (n_2 - n_1)/R_\perp$, which are in meridians at right angles to each other (Fig. 122). There are two *kinds* of images of a point-light source, therefore; and these are not pointlike images in the usual sense but *lines*, called *focal lines* (also called *Sturm's lines*). These focal lines are always at right

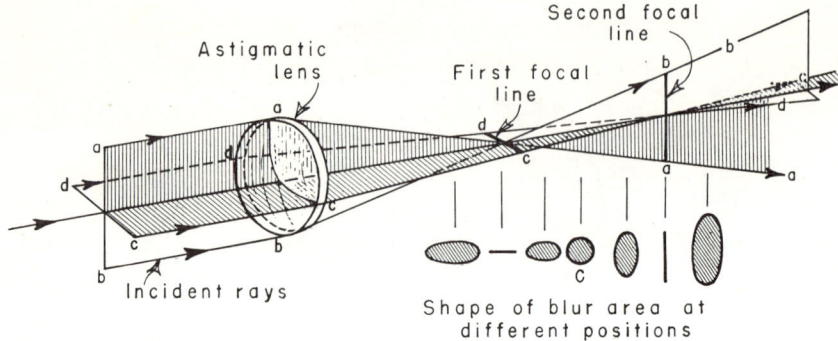

FIGURE 123. Rays used to illustrate the conoid of Sturm and the nature of astigmatic imagery.

angles to each other, and in the symmetric case they are also perpendicular to an optical axis of the surface. The distances of these lines from the surface obey the same formulas for the positions of image and object that were developed for the refraction of a single spherical surface.

The basic characteristic of regular astigmatic imagery can be demonstrated easily by the use of a small flashlight as a point-light source in a dark room. Two ophthalmic lenses taken from the trial set—say, a +5.00-diopter sphere and a +3.00-diopter cylinder—can be held together in one hand (or, better, held in a support such as that on a lens bench). If this lens combination is held at least 1 meter from the flashlight, the focal-line images, one and then the other, can be focused on a sheet of paper held as a screen at the proper distances beyond the lens (Fig. 123). In this simple experiment we can verify the following:

1. The two focal lines are at right angles to each other.

2. The distances of the two focal lines from the lens (if the light source is at a considerable distance from the lens) correspond to the image focal distances produced by the two powers of the lens combination, +8.00 diopters and +5.00 diopters. The distance of the first line from the lens is about 12.5 cm, and that of the second is about 20 cm.

3. The distance separating the two lines is the difference in the focal lengths $(f_o - f_\perp)$ corresponding to the powers in the

two astigmatic meridians. This separation is referred to as the *astigmatic interval* or *Sturm's interval*. However, the astigmatism itself is always defined in ophthalmic optics as the difference between the two dioptric powers—that is, $D_o - D_\perp$ or $(1/f_o - 1/f_\perp)$.

4. Between the two focal lines, the shape of the blurred area of light falling on the screen varies in a characteristic manner. Beginning with the first focal line, the shape of this blurred area changes to an ellipse, to a circle, to an ellipse again, and finally the other focal line (Fig. 123). In an astigmatic eye the nature of the blur disk falling on the retina depends upon the relative positions of the two focal lines with respect to the retina. Furthermore, in general, the larger the blur area, the lower is the visual acuity of the eye.

5. The position of the screen when it is held so that the blur is circular (position $C$ in Fig. 123) corresponds to the *spherical equivalent* of the astigmatic combination; that is, it corresponds to the smallest blur image. This position lies not at the midpoint between the two lines but nearer the first focal line. The distance of the circular spot from the lens corresponds to the reciprocal of the average of the powers in the two meridians. That is, the dioptric *spherical equivalent*, $S$ diopters, is given by $S = \frac{1}{2}(D_o + D_\perp)$. Thus the position of the screen for the circular image is the position corresponding to the smallest blur, or the *circle of least confusion*, obtained when $f_s = 2/(D_o + D_\perp)$.

6. The lengths of the focal lines and, correspondingly, the dimensions of the blurred spots between them depend upon the size of the aperture (the size of the lens in Fig. 123) as well as upon the two focal lengths.

All rays from a given object point that are incident on and emerge from the lens pass through *both* the focal lines (Fig. 123). The light rays from a given object point incident on the lens are restricted to those which can pass through the aperture so that the totality of incident rays actually forms a more or less circular cone, with an apex at the object point and a base at the aperture. The corresponding rays emerging from the astigmatic system also are enveloped by a surface that is more complicated than the simple incident cone. From the point of view of solid geometry, however, this shape is conoid; in ophthalmic optics it is frequently referred to as *Sturm's conoid*. It can be seen easily in a suitably arranged optical smoke box.

In the eye the degree of astigmatism is relatively small compared to the total spherical power of the eye, 59 diopters. The greater part of the measured astigmatism is usually caused by a small toroidal characteristic of the surface of the cornea, but part of the astigmatism may also arise from the two surfaces of the crystalline lens.* Spherical refractive errors can be caused either by spherical errors in curvature of the various refractive surfaces of the eye or by an error in axial length of the eye, or by both. Astigmatism can be caused *only* by errors in curvature, and the presence of astigmatism always means that there are effectively two powers at meridians perpendicular to each other.

The addition of a dioptric spherical power, $D_s$, to an astigmatic system does not change the astigmatism, for the new powers in the two meridians are

$$D_o' = D_s + D_o,$$
$$D' = D_s + D_\perp,$$

and

$$D_o' - D' = D_o - D_\perp.$$

Suppose an eye is completely myopic, which means that both posterior focal (Sturm) lines are in front of the retina (Fig. 124).

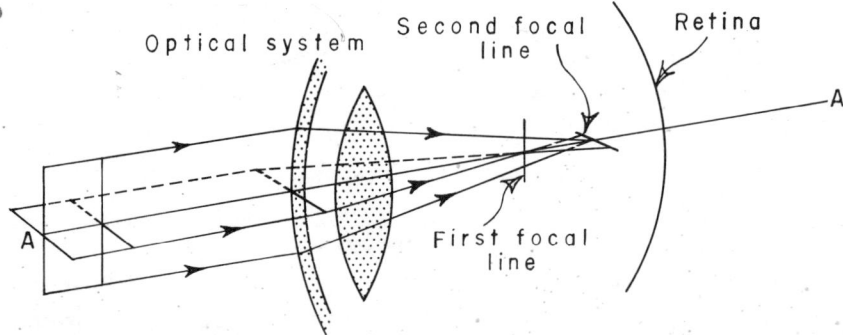

FIGURE 124. The positions of the posterior focal lines in a myopic and myopic-astigmatic eye.

---

*Small astigmatic errors may also arise as a result of lack of centering of the optical elements of the eye.

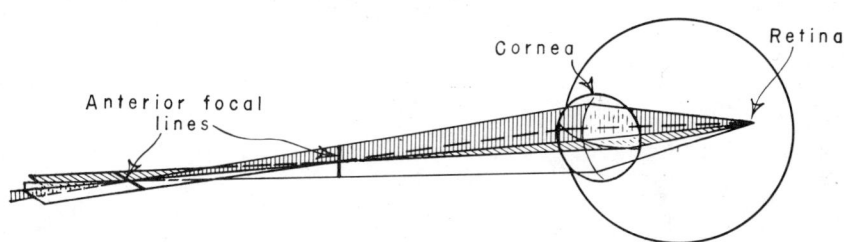

FIGURE 125. The anterior focal lines that are conjugate to the retina of a myopic-astigmatic eye.

If spherical lenses of increasing minus power are placed successively before the eye, then both lines are moved toward the retina: The second focal line falls on the retina, then the first focal line does so, and after that there is an increasing blur of the retinal image. If a corrective lens of both minus spherical power and minus astigmatic power is placed before the eye, with the astigmatic power at the proper meridian, both focal lines can be made to collapse at the retina and the resultant imagery becomes essentially stigmatic.

Inversely, we can trace rays from the retina out through the dioptric system and thus locate the anterior focal lines that are conjugate to the retina of this same myopic and astigmatic eye. These are illustrated schematically in Figure 125. Each of these anterior focal lines is conjugate to the retina and each situated in accord with the refractive powers in the two principal meridians. Just as in the correction of spherical refractive errors, when the refractive error of such an astigmatic eye is corrected with an ophthalmic lens the two focal lines of that lens correction must coincide in position and orientation with the two focal lines that are conjugate to the retina. Both the focal lines that are conjugate to the retina for a completely hyperopic astigmatic eye are located behind the retina.

The vision of persons who have astigmatic errors varies considerably with the degree of the astigmatism, with the refractive powers in the two astigmatic meridians, and with the type of details of the objects observed. In the example of the myopic eye (Fig. 124), in which the focal lines are vertical and horizontal with the horizontal nearer the retina, horizontal lines in an object

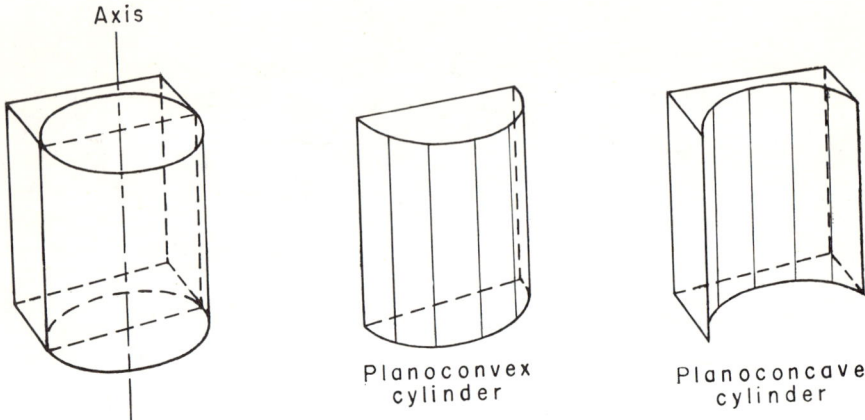

FIGURE 126. Appearances of true geometric cylinders.

produce sharper retinal images than do vertical lines. In the fan or astigmatic dial chart, which consists of radiating black lines, the horizontal lines appear darker (sharper) than the vertical ones because the images of the latter are more out of focus. This follows because the images of all object points on a line parallel to a given focal line converge to that focal line.

Clinically, the refractive error is determined by the use of trial-case lenses consisting of a series of converging and diverging spherical lenses and a series of lenses called "cylindric," likewise of converging and diverging power. The powers of these lenses when used together are additive. The cylindric lenses in some trial-case sets are indeed portions of true geometric cylinders (Fig. 126). Geometrically, such cylinders are generated when a line is rotated about a geometric *axis*, the line and axis being parallel. From this fact has evolved the custom of specifying the orientation of a cylindric lens by the axis of the lens. This type of cylinder has its dioptric power in only one meridian, and that meridian is *at right angles to the axis*. The power in the meridian of the axis is zero, since in that meridian the lens is a plane parallel (Fig. 127). The focal line corresponding to the meridian of cylindric power, $C$ diopters, is *parallel* to the *axis* of the cylinder and at the distance $f = 1/C$, according to the usual lens designation. The second focal line is at $S$.

## Astigmatism

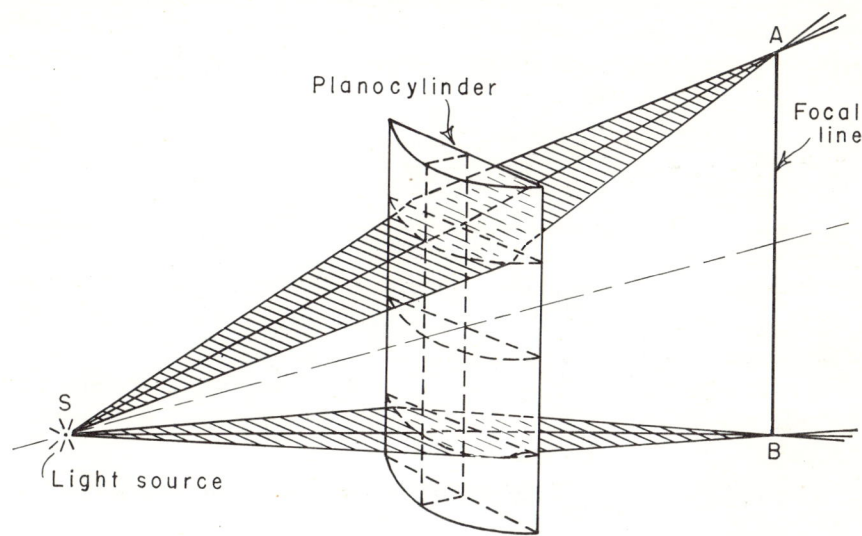

FIGURE 127. Imagery produced by a planoconvex cylindric lens.

It is unfortunate that the orientation of cylinders has come to be specified by the position of the axis rather than by the meridian of power (which is at right angles to the axis), for this designation ofte  ads to confusion. When an ophthalmic cylinder is placed before either eye, the orientation of the cylinder is specified by the angle at which the axis of the cylinder is located, measured in degrees counterclockwise from the horizontal meridian and written as some angle from 0 to 180 arc degrees.

Combinations of spheres and cylinders are used to correct the power errors of an ametropic eye in the two meridians. Since there are both plus and minus cylinders in the trial case, any degree of ametropia can be corrected by either of two methods, depending on which focal line is first brought to the retina by the spherical lens. When the eye has hyperopic astigmatism and both focal lines lie behind the retina, a plus sphere can be used to bring the nearer (first) focal line to the retina, and then a plus cylinder can be added to bring the farther (second) focal line to the retina. Alternatively, if the power of the sphere is such as to bring the farther (second) focal line to the retina, then the other (first) focal line lies in front of the retina. With the use of a minus cylinder this focal line can be brought back to the retina to render the eye stigmatic. In both procedures the two power errors, $D_o$ and $D_\perp$, are corrected by the lenses.

*Examples:*      *Prescription*      *ComponentPowers*

(1)   $+2.00$ sph. $\circ$ $+1.00$ cyl. $\times$ 90°    $\begin{cases} D_{x90} = +3.00 \\ D_{x180} = +2.00 \end{cases}$

or   $+3.00$ sph. $\circ$ $-1.00$ cyl. $\times$ 180°

(2)   $+3.00$ sph. $\circ$ $-4.00$ cyl. $\times$ 60°    $\begin{cases} D_{x60} = -1.00 \\ D_{x150} = +3.00 \end{cases}$

or   $-1.00$ sph. $\circ$ $+4.00$ cyl. $\times$ 150°

In these examples it must be remembered that the actual dioptric effects are in the meridians at right angles to the stated axes. On a fan target those lines appear sharper (blacker) which are parallel to the focal line nearer the retina. The prescription as written can be *transposed* from plus to minus cylinders or vice versa by the following rule for transposition:

     1. Add the powers of the sphere and cylinder algebraically. This gives the new spherical correction.

     2. Use the same cylindric power, but change its sign (from plus to minus or vice versa).

     3. Change the axis of the new cylinder by 90°, stating the result between 0° and 180°.

Obviously, any prescription for the two eyes can be written in four ways.

The spherical equivalent for a given prescription when astigmatism is present is the spherical correction which produces the smallest blurred image on the retina. This can be found by dividing the power of the cylinder by two and adding to this algebraically the power of the spherical lens correction. This procedure follows directly from the discussion under number 5 on page 165, for the power in one meridian would be that of the spherical correction alone, and that for the meridian at right angles would be the power of the spherical correction plus the astigmatic (cylinder) correction. Thus, for a prescription of $+2.00$ sph. $\circ$ $+3.00$ cyl. axis 60°, the spherical equivalent is a $+3.50$-diopter sphere. Without actually correcting the astigmatism of the eye, the spherical equivalent correction gives the next best visual acuity.

For an eye which has both a spherical and a cylindric refractive error, we cannot know *a priori* the basis of this error. In example 2, just given, we do not know whether the eye is basically myopic (axial or refractive, −1.00 diopter) with a +4.00-diopter astigmatic error (because the surface of the cornea, say, is flattened so as to give a lower curvature in the 60° meridian), or whether the eye is basically hyperopic by +3.00 diopters with an astigmatic error of −4.00 diopters (because the surface of the cornea has an increased curvature in the 150° meridian). This observation is important when one tries to ascertain the spherical and astigmatic components of the anisometropia between the two eyes and to predict the possible presence of an aniseikonia.

## 2. DIOPTRIC COMPONENTS OF A CYLINDRIC LENS

Occasionally in ophthalmic optics we encounter the problem of determining the dioptric power of a given cylinder at some meridian other than that of the principal power.

The principal powers of a cylindric surface are $C = (n - 1)/R_c$ in the meridian at right angles to the axis, where $R_c$ is the radius of the cylinder (Fig. 128), and $D_{axis} = 0$ in the meridian of the axis, since the radius of curvature in the meridian parallel to the axis of the cylinder is infinite. Now, a plane $M$ passed through this cylinder at an inclined angle, $\theta$, with respect to the meridian of cylindric power intersects the cylindric surface in a curve. At a point on the *optical* axis, this curve (an ellipse) can be approximated by a circle of radius, $R_\theta$, which is greater than the radius of the cylinder, $R_c$. Correspondingly, the surface at this point has a dioptric power in this meridian, $\theta$, of $D_\theta = (n - 1)/R_\theta$. The mathematics involved in the derivation of the formula for this power in terms of the radius $R_c$ and the angle $\theta$ is too complicated to be given in this discussion. However, it can be shown that the (paraxial) dioptric power of the cylindric surface in the meridian designated by the *axis* angle, $\phi$ (at right angles to $\theta$), is $D_{\times\phi} = C \sin^2\phi$. At right angles the power would be

$$D_{\times(\phi + 90°)} = C \cos^2\phi. \quad \ldots (68)$$

Occasionally a problem is encountered in which a cylinder of power $C$ is placed at an oblique axis, $\phi$, and we need to know the dioptric powers of the cylinder in the horizontal meridian (axis 90°) and in the vertical meridian (axis 180°)—for example,

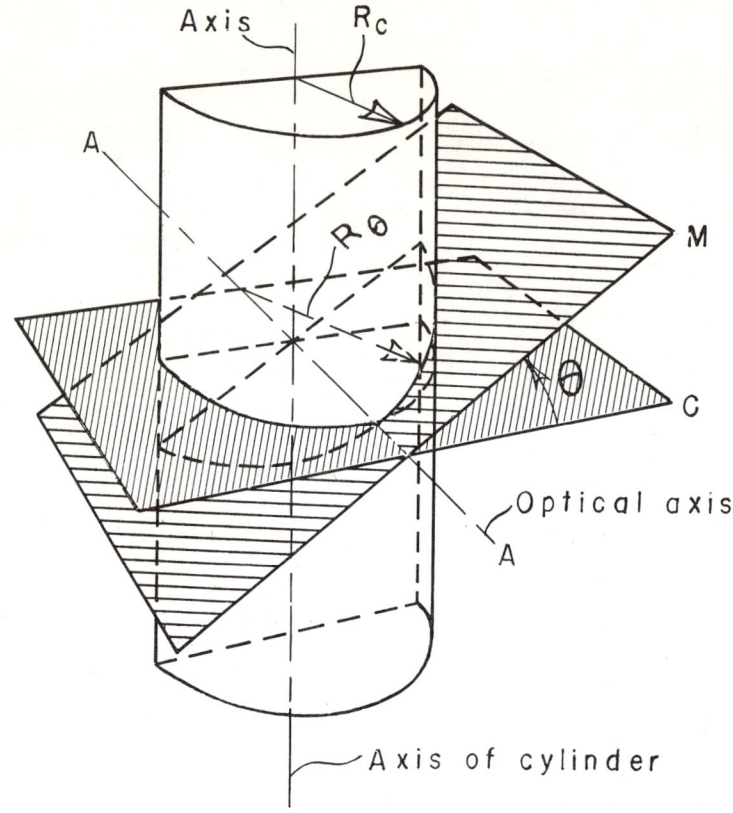

FIGURE 128. The curvature of a planoconvex cylindric surface in an oblique meridian.

in order to be able to determine (by one method) the prismatic effect in these meridians. We have, then,

$$D_{x90} = C \sin^2 \phi \quad \text{and} \quad D_{x180} = C \cos^2\phi, \quad \ldots (69)$$

for the horizontal and vertical meridians, respectively.

*Example:* Suppose a +3.00-diopter cylinder is placed at axis 30° before the eye. What are the powers of that cylinder in the horizontal meridian (axis 90°) and in the vertical meridian (axis 180°)? We have
$D_{x90} = (+3.00) \sin^2 30° = (+3.00) (.50)^2 = (3.00) (.25)$
$$= +0.75 \text{ diopter,}$$
and
$D_{x180} = (+3.00) \cos^2 30° = (+3.00) (.866)^2 = (3.00) (.75)$
$$= +2.25 \text{ diopters.}$$
It should be noted that the sum of the powers of the horizontal ($\times$ 90)

and vertical ($\times 180$) components is equal to the power of the cylinder ($C$) itself.

Approximations to these formulas have been suggested which may be of value clinically. In one,[1] for example, the power of the cylinder ($C$ diopters) is multiplied by a ratio of the pertinent axis angles involved, in the following manner:

$$D_{\times 90} = C \frac{\text{axis angle}°}{90°}, \text{ and}$$

$$D_{\times 180} = C \frac{90° - \text{axis angle}°}{90°}.$$

...(70)

By these formulas the solution of the above example is $D_{\times 90} = 1.00$ diopter, instead of 0.75; and $D_{\times 180} = 2.00$ diopters, instead of 2.25.

## 3. OBLIQUELY CROSSED CYLINDERS

Whenever two cylindric (astigmatic) lenses are combined so that the axes of the two deviate from each other by an angle, $\gamma$, the resultant powers are equivalent to two other cylinders at right angles to each other, and the combination is equivalent to a spherical lens together with a cylindric lens at some particular axis.

It has been shown[2] that the problem of finding the resultant sphere and cylinder can be solved by graphic (and of course analytic) methods, by treating the component powers of the two cylinders as vectors in a vector diagram, but in so doing, the vectors representing the cylindric powers are directed at angles

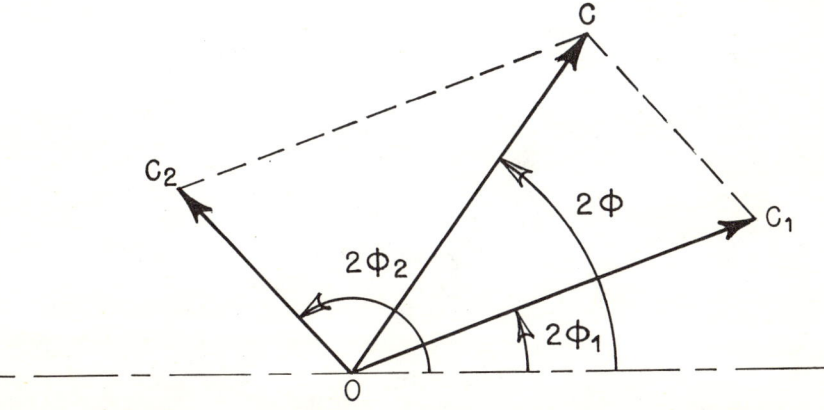

FIGURE 129. Method of representing graphically the resultant of two planocylindric lenses with axes at different angles.

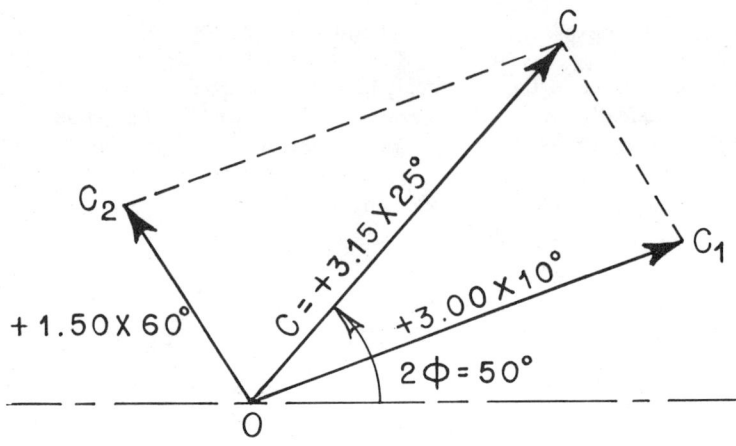

FIGURE 130. The resultant spherocylinder equivalent of two planocylindric lenses placed with axes crossed obliquely.

*twice* the actual angles of orientation before the eye. In Figure 129 is shown such a graphic plot, given a cylinder of power $C_1$ at axis $\phi_1$ and a cylinder of power $C_2$ at axis $\phi_2$. In this kind of graph, cylinders having plus powers are plotted above the line and those having minus powers are plotted below. The angles plotted are always kept between 0 and 180°. Thus $C_1$ is plotted at the angle $2\phi_1$, and $C_2$ is plotted at the angle $2\phi_2$. The parallelogram is then completed, and the resultant vector, $OC$, is drawn. The length of this resultant specifies the power of the equivalent cylindric part of the combination. Its angle, $2\phi$, can be measured with a protractor; and the angle $\phi$, derived from it, indicates the axis of the resultant cylinder. The power of the spherical part, $S$, of the combination must be found from the formula

$$S = \tfrac{1}{2}[C_1 + C_2 - C]. \qquad \ldots (71)$$

*Example:* Suppose a cylinder of power $C_1 = +3.00$ diopters at axis 10° is combined with a cylinder $C_2 = +1.50$ diopters at axis 60°. What is the spherocylinder equivalent? Figure 130 shows how these data may be plotted by use of lines (2 cm = 1 diopter of power) directed at angles which are twice the given axis angles. The parallelogram is completed, and the resultant line, $OC$, is drawn. This resultant is then measured and found to represent a resultant cylinder of +3.15 diopters. The angle is then measured with a protractor and found to be 50°, so the angle of the resultant cylinder, $\phi$, is 25°. The spherical part of the combination is then found from $S = \tfrac{1}{2}(3.00 + 1.50 - 3.15) = +0.67$ diopter. Hence the spherocylinder equivalent of the two cylinders crossed at oblique axes would be a +0.67-diopter sphere combined with a +3.15-diopter cylinder at axis 25 degrees.

The problem of cylinders with axes crossed obliquely occurs also in the manufacture of certain aniseikonic lenses in which bitoric lenses (both sides of which are toric surfaces) must be ground. Such lenses are designed to provide different magnifications in the two meridians which do not coincide with the axis of the astigmatic correction.

If the powers of the two cylinders are the same, the axis of the equivalent cylinder is midway between the axes of the two. If the axes of these two equal cylinders are at right angles, however, the result is a lens with spherical power only.

One of the purposes of including the subject of crossed cylinders at oblique axes in this discussion is to point out several factors in the determination of the cylindric correction of astigmatism. We may consider the astigmatic eye itself as one cylinder, the trial-case cylinder, with power of opposite sign, as the second. The aim of the refractionist is to find the power and axis of the correcting cylinder which will offset exactly the cylindric power error of the eye. By the method of diagramming presented above for crossed cylinders at an oblique axis, the lines representing the cylindric powers of eye and lens would be the same length but in opposite directions from the origin, 0. Any residual spherocylindric lens power of the combination would be zero. If the axis of the correcting cylinder was correctly determined but the power was incorrect, the residual cylindric power would still be at the same axis. The lines on the fan dial would appear darker in the meridian of this axis.

If the power of the cylinder was correct but the axis was incorrect, there would be a residual uncorrected cylindric power whose

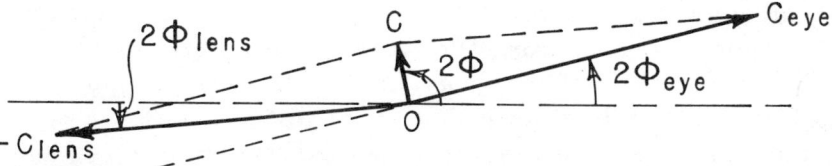

FIGURE 131. The resultant astigmatism in the case of two cylinders of equal but opposite power, with a small angle between the two axes.

axis would be very different from the axis of astigmatism of the eye—indeed, nearly 45° away from that axis. This can be seen from inspection of Figure 131. There the power of the eye-cylinder is plotted as plus and that of the corrective cylinder as minus (hence below the reference line). The two are shown as being slightly off axis. The line indicating the residual cylinder (the uncorrected astigmatism) would correspond to the resultant of the completed parallelogram. This resultant line is approximately 90° from the axis of the eye-cylinder; therefore the axis of the residual cylinder would be about 45°. Consequently, rocking the axis of the corrective cylinder from one side to the other of the true axis of the ocular astigmatism causes any small uncorrected astigmatism to jump 45° to one side and the other of the true axis. Under this procedure the lines on the astigmatic dial (fan chart) would alternately grow darker at 45° on one side, then on the other, of the true axis of astigmatism.

## 4. THE CROSS-CYLINDER TEST LENS (JACKSON)

A cross-cylinder test lens is frequently used to refine the correction for the astigmatism of the eye. This lens consists of two cylinders of equal powers but of opposite sign with axes at right angles to each other. One that is commonly used consists of a +0.25-diopter cylinder at one axis combined with a −0.25-diopter cylinder at an axis 0° from the first. In prescription form, such a test lens is described as a spherocylinder lens—either +0.25 sphere ◯ −0.50 cylinder, or −0.25 sphere ◯ +0.50 cylinder. The power of the spherical equivalent of this lens is zero. When used before the eye, therefore, the cross cylinder has little effect on the spherical part of the refractive error of the eye. When the axes of this lens coincide with the axes of astigmatism of the eye, the separation of the Sturm (focal) lines is either increased or decreased (tending toward a correction of the astigmatism), depending upon the orientation of the axes of the cross-cylinder lens.

For a discussion of the general use of the cross cylinder as a tool to find the best correction of astigmatism of the eye, the reader is referred to texts on refraction procedures.

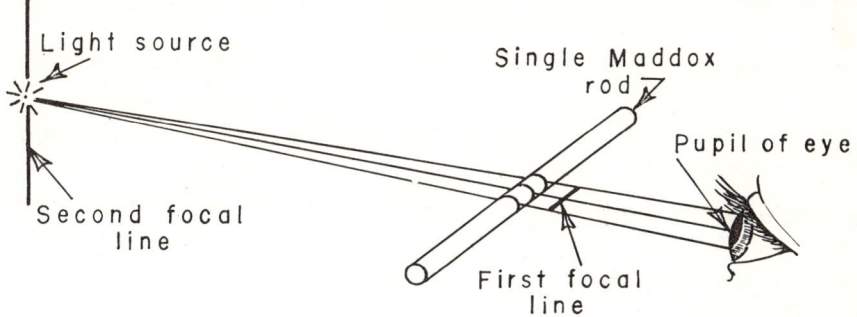

FIGURE 132. The focal lines produced by the Maddox rod.

## 5. THE MADDOX ROD

The Maddox rod is a device which, placed before one eye, conveniently prevents binocular vision and at the same time permits one to estimate subjectively the direction of the visual axis of that eye relative to that of the other eye. This device consists of a row of small glass cylinders or rods (Fig. 132). Each of these rods acts as a cylindric lens with very high power so that the first focal line of a distant point of light is parallel to the axis of each rod and is *very* close to the rod. The second focal line, of course, is very far away—usually at the distance of the point source of light. The latter focal lines, because their edges are not sharp, are usually localized only indefinitely by the eye behind the rods. The first focal lines are so close to the eye that their images in the eye are extremely blurred and are so diffuse that they are not seen at all. The second focal lines from a series of rods are seen as a continuous, long, sharp line at *right angles* to the axes of the rods.

## 6. THE SCISSORS EFFECT OF CYLINDRIC LENSES

When we look through a cylindric lens at a chart consisting of vertical and horizontal lines, the images of the lines appear to remain vertical or horizontal only if the axis of the cylinder is vertical or horizontal. If the axis is oblique, then the images of the lines no longer appear at right angles, but the former verticals and horizontals appear slightly rotated in directions opposite each other, making the included angle either acute or obtuse (Fig. 133).

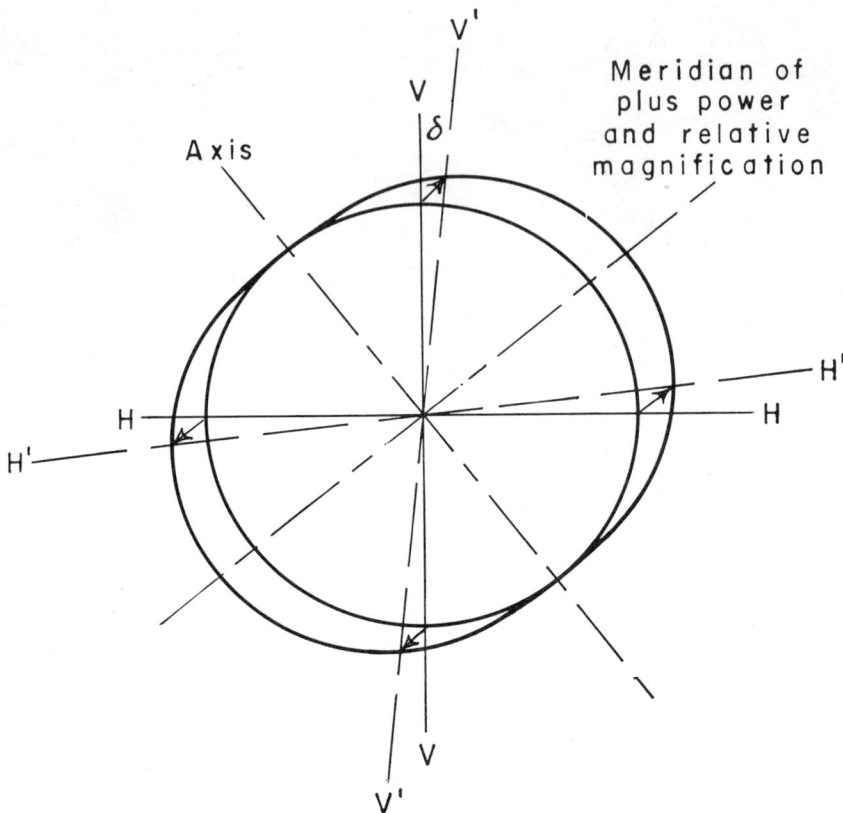

FIGURE 133. The rotary deviation of the images of vertical and horizontal lines produced by a cylindric lens placed at an oblique axis.

This effect is due to the fact that the angular magnification is different in the two principal meridians of the cylinder (see Chapter XII, section 1). In general, the magnification is greater in the meridian of higher plus power. The image may actually be minified in the meridian in which the power is minus.

The figure illustrates the optical effect produced when the magnification introduced by a cylinder with plus power is at an oblique axis. The magnification occurs only in the direction of the power meridian, and the displacement of the image of any point in that direction is proportional both to the magnification and to the angular distance of the point from the axis of the cylinder. The circle in the figure represents object points equidistant from the optical axis of the lens. The images of these points are displaced in the meridian of magnification so as to lie on the ellipse. Thus points on the objective vertical line $V$ and the objective horizontal line $H$ are displaced to positions on lines $V'$ and $H'$, respectively. Thereby, the images of both the vertical and the horizontal lines are rotated slightly, in directions opposite each other, toward the meridian of magnification. This effect is frequently referred to as a *scissors effect*.* It occurs in many problems in which astigmatic lenses at oblique axes are used.

## EXERCISES

1. A toric glass surface has radii of curvatures of 133 and 83.3 mm. What are the powers of this surface? Assume that the index of refraction of the glass is 1.5. What is the astigmatism?

2. A lens has a front toric surface with powers of 8 and 10 diopters. The back surface is a −6.00-diopter sphere. What is the spherocylinder power of the lens (neglecting the thickness)?

3. An astigmatic correction lens has the power +3.00 sphere ○ +3.00 cylinder. At what distances are the focal lines from the lens for a distant point-light source object? What is the separation of the lines—the astigmatic interval? What is the spherical equivalent of the lens? At what distance from the lens will the circle of least confusion be found?

4. If a +3.00-diopter sphere is added to the lens in problem 3, at what distances from the combination are the focal lines to be found? What is the astigmatic interval?

5. Calculate the front surface (toric) powers in the two meridians of the astigmatic lens +3.00 sphere ○ −4.00 cylinder (vertex powers), if the back curve is (a) −8.00 diopters and (b) −4.00 diopters. Take the thickness in both instances as 3.5 mm. Hint: Solve equation 55 for $D_1$.

6. What is the spherical equivalent in each of the following prescriptions?
    (a) +2.00 sph. ○ +3.00 cyl. × 60°
    (b) −5.00 sph. ○ −3.00 cyl. × 180°
    (c) +2.00 sph. ○ −4.00 cyl. × 90°
    (d) −3.00 sph. ○ +4.00 cyl. × 90°

---

*This effect must be distinguished from a different kind of scissors effect in retinoscopy, in which a small amount of irregular astigmatism which may be present disturbs the shape of the retinal reflex.

7. Determine the horizontal and vertical component powers of each of the following astigmatic lenses:
    (a) +3.00 cyl. × 45°
    (b) +4.00 cyl. × 60°
    (c) +2.50 cyl. × 120°
    (d) +3.00 cyl. × 150°
    (e) −3.00 cyl. × 30°
8. If each of the lenses in problem 7 is decentered 8 mm "down" and 5 mm "in," what is the prismatic deviation (prism power and base direction) produced in each case?
9. Transpose each of the following:
    (a) +4.00 sph. ◯ −2.00 cyl. × 33°
    (b) −5.00 sph. ◯ −3.00 cyl. × 160°
    (c) −4.00 sph. ◯ +3.00 cyl. × 90°
    (d) +2.00 sph. ◯ −3.00 cyl. × 180°
    (e) −2.50 cyl. × 40°
    (f) +4.00 cyl. × 160°

## REFERENCES

1. SNYDACKER, DANIEL, and NEWELL, F. W.: *Refraction: A Home Study Course.* Rochester, Minn., American Academy of Ophthalmology and Otolaryngology, Section on Instruction, 1964, p. 61.
2. SOUTHALL, JAMES P. C.: *Mirrors, Prisms and Lenses. A Text-Book of Geometrical Optics*, 3rd ed. New York, Macmillan, 1940, pp. 320-326. Reprinted 1964, New York, Dover.

Chapter XI

# ABERRATIONS OF SPHERICAL LENS SYSTEMS

Most lens systems have certain inherent characteristics that tend to cause some deterioration of the image, thus preventing it from being truly stigmatic. There are five of these deteriorating characteristics, called *aberrations*, and several or all of them may be acting at the same time. The aberrations affect the image in the human eye and also, to some extent, the images formed by the ophthalmic lenses used to correct refractive errors of the eye. The effects of the aberrations of the ophthalmic lens may differ somewhat from those found in photographic systems, for example, because the eye is free to rotate behind the corrective lens. As the eye turns to fixate the images of obliquely located objects, entirely different portions of the lens are used, since the pupil limits the size of the effective pencil of light passing through the lens.

The several aberrations are classified according to the particular manner in which each effects a deterioration of the image. These usually are given as follows.

## SPHERICAL ABERRATION

Spherical aberration occurs because, in the usual optical system, the rays which enter the system away from the axis (the more marginal rays) focus nearer the lens than do those which enter close to the axis (Fig. 134). This aberration tends to cause some blurring of the image of a point object *on the axis* of the lens system, especially in those systems with large apertures. The geometric surface which envelops the totality of rays emerging is called, just as in the case of the mirror, a *caustic surface*. Only the caustic curve in one plane can be shown in the figure. The blurring of the image caused by spherical aberration is symmetric about the axis of the lens system; and obviously, the smaller the aperture,

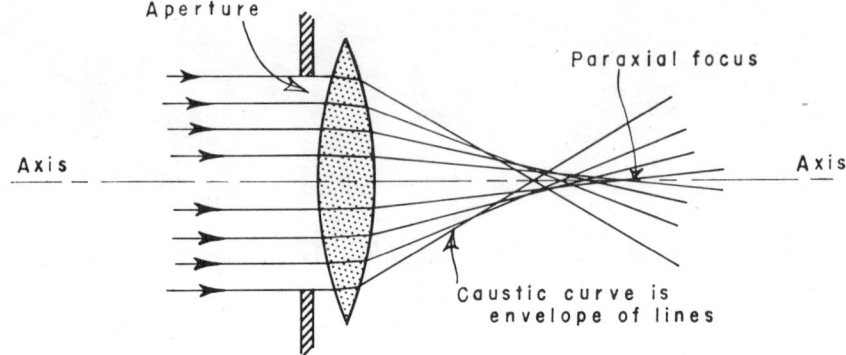

FIGURE 134. Positive spherical aberration of a lens.

the smaller will be the blurring effect. In Figure 134, the spherical aberration is said to be positive. Some complex systems may have negative aberration, and in these the extra-axial rays are focused farther from the system than are the axial rays. In certain optical systems, of which the normal eye may be one, the spherical aberration may vary from positive to slightly negative for rays that pass through the pupil at increasing distances from the axis.

The spherical aberration of a single lens varies considerably with the shape (the coflexure) of the lens. A meniscus plus lens with concave surface *toward* the object has an especially large aberration; whereas the biconvex lens, with which the surface toward the object has a slightly greater curvature, has less spherical aberration.

The spherical aberration of a single lens, like that of a mirror, can be eliminated if one surface can be ground and polished to a particular *aspheric* curve. Insofar as ophthalmic lenses are concerned, the advent of plastic materials will make aspheric surfaces more readily obtainable. These surfaces may be especially important in corrective lenses of high power, such as cataract lenses and lenses for subnormal vision.[6]

## 2. CHROMATIC ABERRATION

Chromatic aberration, as the name implies, results from the fact that the index of refraction varies with the wavelength of

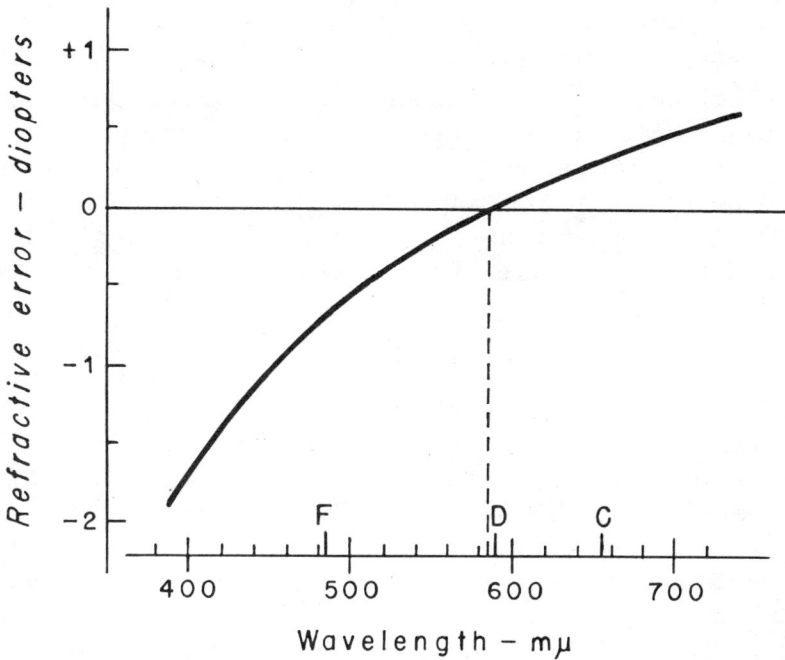

FIGURE 135. The refractive error of the eye due to chromatic aberration associated with monochromatic light of different wavelengths.

light. Since the index of refraction for red light is less than that for blue, the focal length of an ordinary lens is longer for red light than for blue: Blue rays focus nearer the lens than do red rays. Chromatic aberration also tends to blur the images if white light is used.

The human eye has chromatic aberration, but except under laboratory conditions its effect is rarely seen. The equivalent refractive error of the human eye for monochromatic light of various wavelengths is shown in Figure 135.[1] The relationship, in respect to the eye, is not a linear one, but the error becomes increasingly myopic toward the blue end of the visible spectrum. The eye tends to focus for light of a wavelength near that corresponding to the D (Fraunhofer) line.[3] The total chromatic aberration between the F and C lines of the spectrum is about 0.9 diopter. A cobalt glass (which transmits light only from the blue and red ends of the spectrum) may be useful in determining refractive errors. In fixating a point source of light through the glass, the overaccommodated eye (myopic eye) sees a blue ring of light surrounding a red spot, whereas an underaccommodated eye (uncorrected hyperopic eye) sees a red ring of light surrounding a blue spot.

Chromatic aberration can be somewhat reduced in lenses if the glass used has a small chromatic dispersion, and it can be eliminated in doublet systems by combining several lenses made of glasses of different indices of refraction and different $\nu$ (nu) dispersion values.

### 3. COMA

Coma is essentially a type of spherical aberration affecting the light from object points which lie off axis (Fig. 136)—that is, a type of spherical aberration of pencils of light at oblique incidence to the lens. The blurring caused by coma differs from that of spherical aberration—which arises from an axial object point—in that there is no center or axis of symmetry. The shape of the blurred image on a screen resembles a raindrop or a comet (hence the name). Chromatic aberration causes further deterioration of the image of an obliquely located point source so that the blur spot has asymmetric colored fringes.

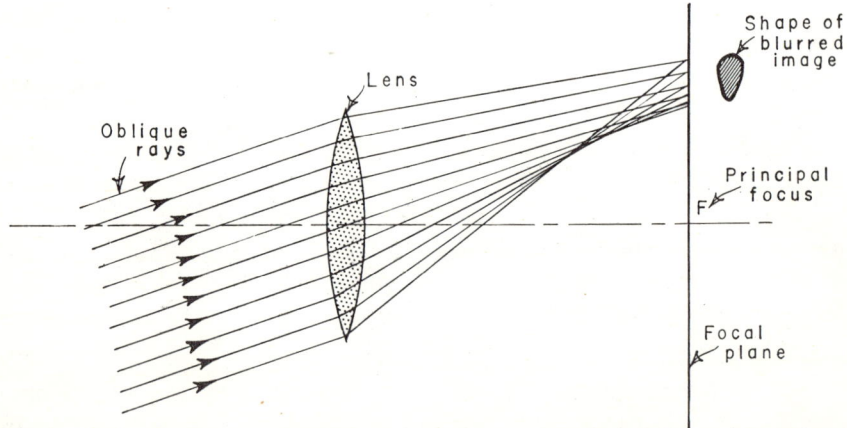

FIGURE 136. The aberration called *coma*, showing the type of blur of the image of an extra-axial object point.

## 4. ASTIGMATISM OF OBLIQUE INCIDENCE

When a pencil or beam of light is incident to the surface of a lens at a large angle of oblique incidence as measured from the optic axis, the emerging refracted pencil fails to focus as a stigmatic image (even apart from the effect of coma and chromatic aberration) and is actually astigmatic. This aberration, *astigmatism of oblique incidence*, is illustrated by exaggerated perspective diagrams in Figures 137 and 138. This astigmatism results from the fact that the spherical wave front from the object point meets the front lens surface at an oblique angle in such a manner that, effectively, the surface has two radii of curvature, and hence has two powers. The *tangential* (or primary) focal line is perpendicular to the plane formed by the optical axis of the lens and the incident chief ray of the pencil. The *radial* (secondary, and frequently called *sagittal*) focal line lies in that plane. The separation of the two focal lines

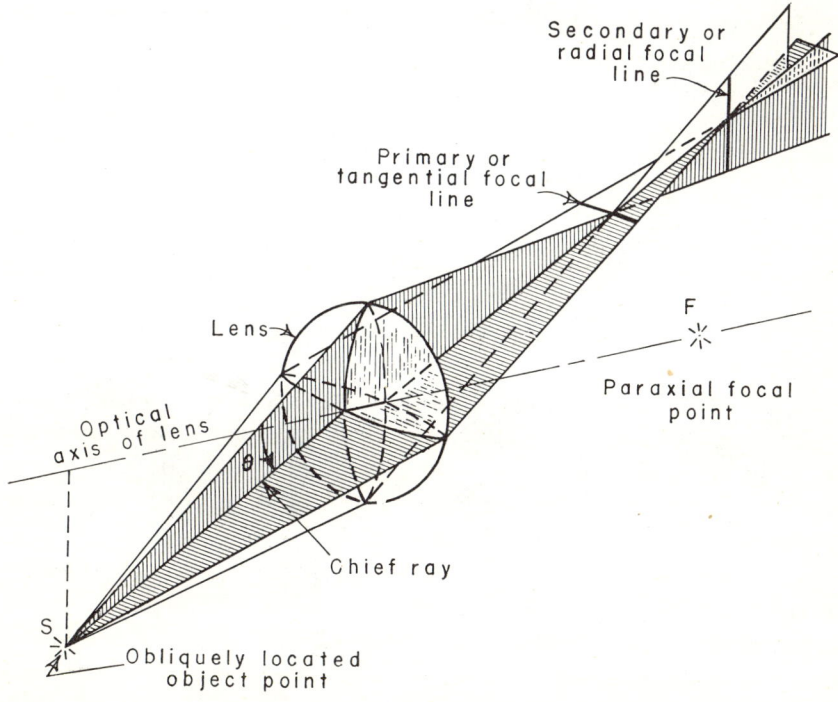

FIGURE 137. The astigmatism of rays of light incident to the lens at an oblique angle.

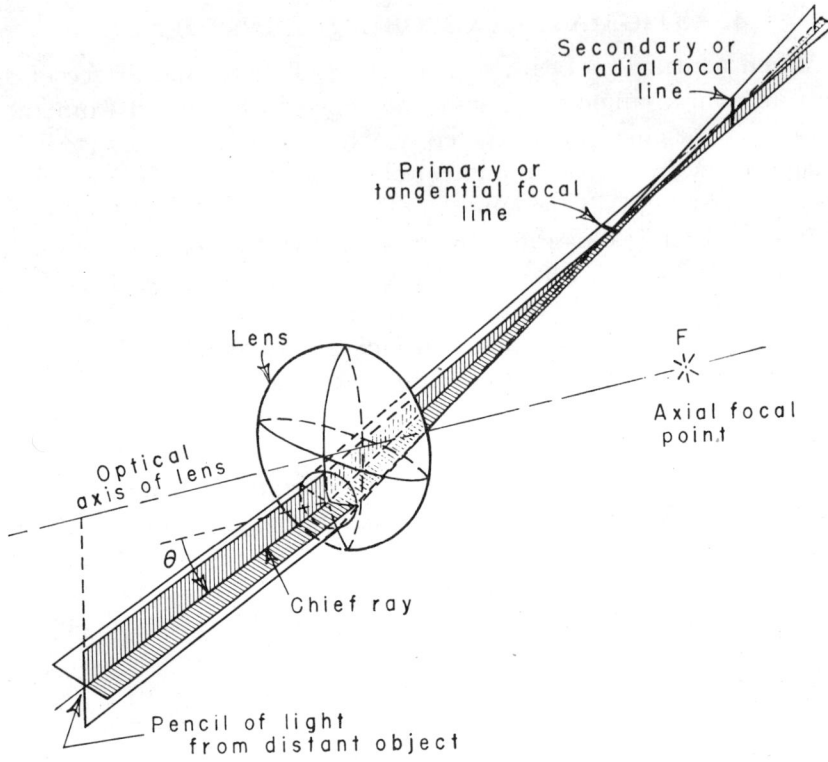

FIGURE 138. The astigmatism of rays at oblique incidence limited to a small area of the lens.

(that is, the astigmatic interval) is independent of the size of the incident pencil (or aperture of the lens). The lengths of the focal lines, however, do depend upon the size of the aperture. The positions of these focal lines—in terms of the surface powers, angle of obliquity, and distance of object—can be described mathematically, but the relationships will be omitted from this discussion.

In a given lens system the positions of these two focal lines can be measured for a wide range of oblique angles. (Usually the measurements are referred to the theoretic focal plane of the lens.) The resulting surfaces for the successive positions of the tangential and the radial focal lines may be represented as shown in Figure 139. The locus of points for which the smallest blur circles will be

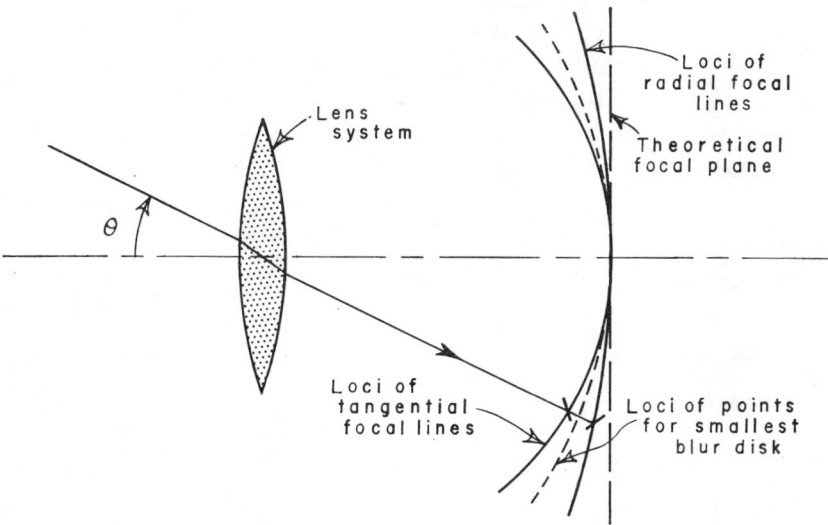

FIGURE 139. Loci of the tangential and radial images of a lens system, illustrating the *curvature of field*.

found lies, of course, between these two surfaces. This intermediate surface determines the *curvature of field* produced by the lens. A photographic lens should be so designed that this surface tends to lie on the plane of the flat film. For the human eye, measurements show that the spherical surface of the retina tends to coincide rather well with the curvature of the field produced by the dioptric system of the eye.

Astigmatism of oblique incidence is a problem in ophthalmic spectacle lenses also, for when the eye is turned to view a lateral object point, the pencil of light as limited by the pupil is incident on the lens surfaces at an oblique angle (Fig. 138). To a certain extent the magnitude of such astigmatism of oblique incidence for an ophthalmic lens can be reduced by judicious selection of the base-curve, the selection depending upon the vertex power of the lens (see Chapter XII, section 4).

## The Astigmatism from Tilted Lenses

When a spherical ophthalmic lens is tilted about an axis in the plane of the lens—for example, about a horizontal axis—the image of a point-light source becomes slightly astigmatic for the same reasons that rays at oblique incidence are astigmatic. Approximately, the amount of astigmatism produced is proportional to the square of the tangent of the angle of tilt.[4] Measurements of the spherocylinder powers for biconvex lenses tilted about a *horizontal* axis are given in the following table (in diopters).

POWERS (DIOPTERS)* OF BICONVEX LENSES TILTED ABOUT A HORIZONTAL AXIS

| 0° | 5° | | 10° | | 15° | |
|---|---|---|---|---|---|---|
|  | Sph. | Cyl. ×180° | Sph. | Cyl. ×180° | Sph. | Cyl. ×180° |
| +2.00 | +2.00 | 0.00 | +2.02 | +0.06 | +2.04 | +0.28 |
| +4.00 | +3.97 | +0.11 | +4.02 | +0.19 | +4.16 | +0.41 |
| +8.00 | +7.98 | +0.12 | +8.04 | +0.19 | +8.13 | +0.45 |
| −3.00 | −3.01 | −0.02 | −2.98 | −0.20 | −3.06 | −0.39 |
| −6.00 | −5.97 | −0.18 | −6.00 | −0.42 | −6.17 | −0.69 |

*Error of measurement = ±0.02 diopter.

We note from these measurements that the power in the vertical meridian (× 90) is only little affected by the tilt, but the cylindrical power in the horizontal meridian (× 180) increases markedly with the angle of tilt.

The astigmatism arising from the tilting of ophthalmic lenses is probably not generally a serious problem unless the powers are high and the tilt is great. Frequently the corrective lenses are deliberately tilted (the so-called pantoscopic tilt) to reduce the astigmatism of oblique incidence in the reading position of the eyes. Thus a compromise is about 7.5° of tilt, since the eyes turn down about 15° when reading.

## 5. DISTORTION

The aberration classified as *distortion* is not actually concerned with the sharpness of focus of the image but with the faithfulness with which the pattern of the total image of an extended object is reproduced. For example, the image of a checkerboard pattern formed by a simple optical system may be distorted in that the images of the parallel lines making up the pattern are curves, and the images of the squares are unequal in size or unlike in shape.

The distortion is greatly influenced by the position of the *stop* or aperture which restricts the rays passing through the system.

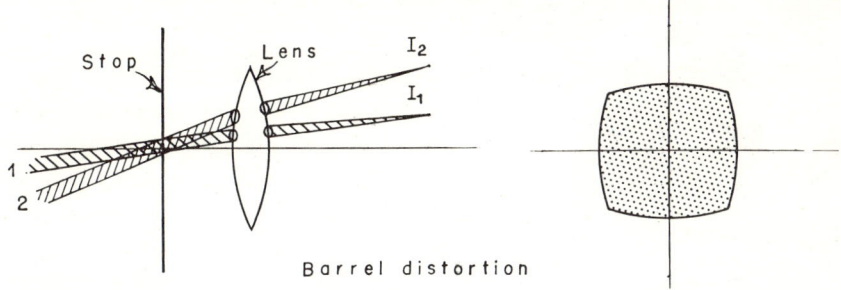

FIGURE 140. Barrel distortion of the image formed by a lens system, associated with decreasing magnification away from the optical axis.

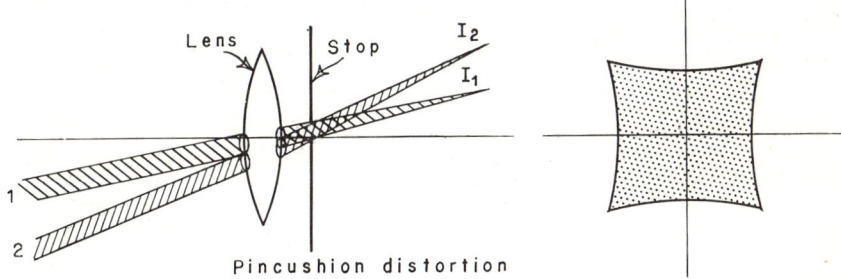

FIGURE 141. Pincushion distortion of the image formed by a lens system, associated with increasing magnification away from the optical axis.

The distortion occurs because the different pencils of light from object points located at different angles from the optical axis are directed to different parts of the lens (Figs. 140 and 141). Whether or not the image is distorted depends upon whether the magnification of the image is constant or whether it varies with distance from the axis of the lens system. In the latter case, two types of distortion for centered systems are usually distinguished, each of which is described according to the shape of the image of a square.

In *barrel* distortion the images of the straight sides of the square are curves bowed outward like the sides of a barrel. This type of distortion indicates that the magnification of the image is not uniform but *decreases* slightly with the distance of the object point from the optical axis (Fig. 140). Since the corners of a square are farther from the center than are the midpoints of its sides, the images of the corners will not be proportionally far enough from the center to produce a square image. The human eye is said to exhibit barrel distortion.

In *pincushion* distortion the images of the straight sides of a square object curve inward as do the edges of an old-fashioned pincushion (Fig. 141). This type of distortion indicates that the magnification of the image is not uniform but *increases* slightly the farther the object point lies from the optical axis. The images of the corners of the square, being the parts farthest from the center, are magnified more than the images of the sides.

In noncentered optical systems the distortion of the image may also not be symmetric about the optical axis, as in the two cases just mentioned. An example is the distortion of the image by a prism, which is due to *asymmetric* magnification (Fig. 60). Since the optical components of the human eye are not coaxial, the distortion of the image on the retina may be of this asymmetric type.

A discussion of distortion of the image of an extended object should also mention the phenomenon of differences in magnification resulting from chromatic aberration. In this effect—called *chromatic difference in magnification*—the red rays of the light from the various parts of the object may produce an image of different size than would the blue rays from the same parts. In the human eye, in which the optical surfaces are not coaxial, the chromatic difference in magnification may vary for different parts of the visual field.

## 6. SPECIAL TOPICS

The remaining portion of this chapter will be given over to brief discussions of three subjects that are important to visual optics but only indirectly related to the matter of aberrations. For the most part, they are factors limiting the performance of the eye as an optical instrument and are imposed by the nature of light and by the characteristics of the optical system and of the mosaic structure of the retina.

### A. Blurred Imagery

When the eye fixates an object point—for example, $A$ in Figure 142—the accommodation is adjusted so that the image $a$ of this point is defined sharply on the retina. The images of all other points in space not at the same distance as the fixation point are

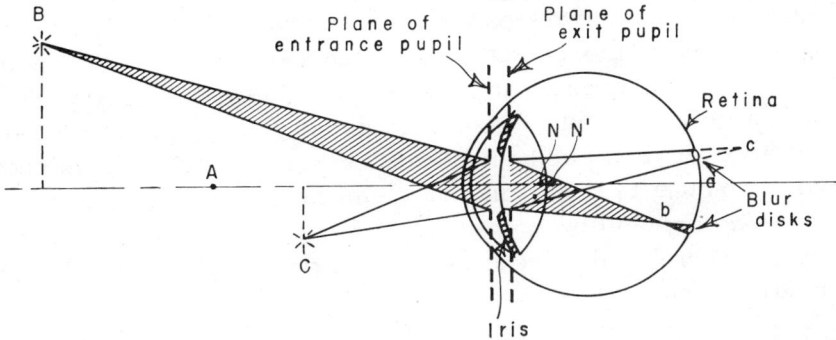

FIGURE 142. Blurring of the retinal images of objects not at a point conjugate to the retina.

less sharp—that is, are out of focus and hence will be more or less blurred disks on the retina. Point $B$, being farther than the fixation point, has its image at $b$, in front of the retina, so there is a small blur disk on the retina. Similarly, point $C$, being nearer than the fixation point, has its image at $c$, behind the retina, the rays being intercepted by the retina to form, again, a small blur disk. The particular pencils of light involved in this imagery are determined by the sizes and positions of the entrance and exit pupils.

The *entrance pupil*, which was discussed on page 138, is the image of the edge of the iris, or of the actual pupil, formed by the corneal refractive system only. It is this pupil which one sees on looking at the eye. The *exit pupil*, similarly, is the image of the pupil formed by the crystalline lens, principally the second surface of the lens. It can be shown that the centers of the entrance and exit pupils are conjugate points of the whole dioptric system of the eye. Hence, a chief ray directed to the center of the entrance pupil emerges from the crystalline lens as if from the center of the exit pupil. Such chief rays lie in the center of the pencils of light from the object point to the image. The centers of the blur disks on the retina correspond to the points of intersection of these central or chief rays.

The angle between the optic axis and the chief incident ray from the object point differs from that between the axis and the corresponding refracted ray to the image. There is, however, a constant ratio between these two angles. The angle between the blurred images of two separated object points is the angle between the centers of the blur circles of the two images. Nodal-point construction is theoretically incorrect for blurred imagery. When the magnification of out-of-focus retinal images is specified, the entrance and exit pupils should be used as the points of reference. In sharp imagery this distinction disappears, since ray constructions by *nodal points* and by *entrance-exit pupil centers* yield the same results. Thus, generally, the entrance and exit pupils are the more useful reference points.

## B. Resolving Power

The smallest angular separation of two points (lines or other similar details) whose images can just be resolved, or can just be discriminated as two (Fig. 143), is called the *minimal angle of resolution*. By long usage this angle is expressed in minutes of arc and under ideal conditions in the normal eye is about 1 minute of arc. The *resolving power* is defined as the reciprocal of this angle, $(1/\alpha)$, when expressed in minutes of arc. Thus, if two points subtending 1 minute of arc can just barely be resolved, the resolving power is 1. If the separation must subtend 2 minutes of arc to be resolved, the resolving power is 0.5, and so forth. *Visual acuity*, which is a more general term for sharpness of vision, depends in large measure upon the resolving power of the eye.

The resolving power depends upon the size of the image blur disk and the dimensions of the receptors in the mosaic structure of the retina of the eye. The contrast threshold and the adaptation level of the retina are also involved. The theory of resolution postulates that for two points of light (stars) to be resolved by the eye, there must be one *less*-stimulated retinal receptor lying between more highly stimulated receptors upon which the images of the two points fall.

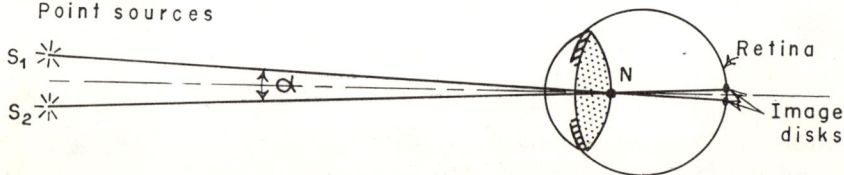

FIGURE 143. Visual angle subtended at the nodal points by two object points.

It is quite often assumed that the limit of optical resolution is set by diffraction and in particular by the size of the central diffraction disk. On the basis of theory we could not expect an angle of resolution, $\alpha$, smaller than

$$\alpha = 1.22\, \lambda/d,$$

in which $\lambda$ is the wavelength of the light and $d$ is the diameter of the aperture (pupil). In this relationship $\alpha$ is in radians; to convert to minutes of arc it must be divided by 0.00029 (1 minute of arc is equal to 0.00029 radian). Using a pupil diameter of 3 mm and the wavelength corresponding to the D spectral line, this minimal angle of resolution is found to be about 0.8 minute of arc.

When the diameter of the pupil increases, the corresponding decrease in the size of the diffraction disk on the retina is offset by an increase in the size of the disk due to the effect of spherical aberration. Experiment indeed shows that visual acuity changes little for pupil sizes greater than about 2 mm in diameter.

## C. Depth of Focus

The size of the blur circle, or preferably the blur disk, on the retina depends upon the extent to which the image is out of focus and upon the size of the aperture or pupil. If the image is out of focus by only a small amount, the blur disk is not much larger than that for the sharpest focus; or at least the difference cannot be discriminated. The degree to which the image can be out of focus without the difference being perceived according to some criterion is a measure of the *depth of focus*.

In Figure 144, suppose the eye is accurately accommodated for the point $S$, a distance $s$ from the eye; the image of $S$ is sharply

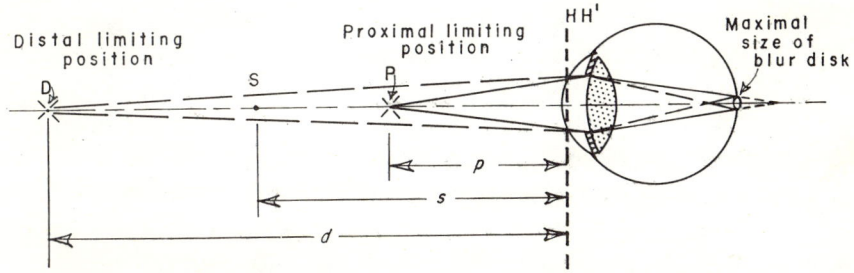

FIGURE 144. Figure illustrating the depth of focus of the eye.

defined on the retina. If now a test point-source of light or a suitable target is moved along the visual axis, nearer or farther from the eye, the image falling on the retina becomes more or less blurred. Any criterion of discrimination would set some maximal permissible size of this blur disk, and this maximal blur disk would correspond to the test target's being at either a *proximal* position $P$ or a *distal* position $D$ (at distances $p$ and $d$, respectively) from the cornea of the eye. The actual spatial distance or interval between the points $P$ and $D$ is called the *depth of field*.

The total depth of focus $T$ is defined as $T = 1/p - 1/d$, which, again, is a difference in vergences. If $p$ and $d$ are expressed in meters, then the depth of focus $T$ is in diopters. Since the depth of focus depends only upon a maximal permissible size of the out-of-focus blur disk, the quantity $T$ is constant, for the same test conditions and same size of pupil, regardless of the distance of the fixation point, $S$, from the eye. That part of the depth of focus given by the interval from sharpest focus to the proximal distance $p$ is $T/2 = 1/p - 1/s$, and that to the distal distance $d$ is $T/2 = 1/s - 1/d$.

The actual depth of field $(d - p)$ changes greatly with the distances of the fixation point, $S$. The distal distance is $d = s/(1 - sT/2)$, and the proximal distance is $p = s/(1 + sT/2)$. The depth of field $(d - p)$ is $Ts^2/[1 - (sT/2)^2]$. We see therefore that the depth of field increases faster than the square of the distance of the fixation point.

*critical dist*

As the fixation point, $S$, moves away from the eye, a critical distance is reached at which the distal point, $D$, is infinitely far away. This follows from the equation above for $d$, for as $s$ approaches $2/T$, the denominator becomes zero, whence $d$ approaches infinity. This critical distance of the fixation point is called the *hyperfocal distance*. With accurate accommodation on a point at this critical distance $s_c$, all objects from the proximal distance $p = s/(1 + sT/2)$ to infinity will be clearly perceived because the size of the blurred images of all the object points will be less than the maximum permissible. Thus the hyperfocal distance $s_c = 2/T$ (meters).

The farthest distance that an object can be seen clearly (the distal limit of the depth of field) by an eye with myopia of $A$ diopters would be $d = 1/(\frac{1}{2}T + A)$.

Several criteria can be used to determine the depth of focus: (1) the discrimination of least perceptible blurring of the image itself; (2) the loss of resolving power or visual acuity, which depends upon the size of the test details used; and (3) the loss of visibility, which varies greatly with contrast of test detail against the background. In ophthalmology, the criteria of least perceptible blur and the loss in visual acuity are both important. A person with a pupil of average diameter (4 mm) has a total depth of focus (at the 50 per cent level of recognition) of about 1.5 diopters, based on visual acuity of print of the 20/30 size.[5] For the least perceptible blur, the total depth of focus for a 3-mm pupil is roughly 1.0 diopter.[2]

The influence of the depth of focus must be taken into consideration in the determination of the near point of accommodation. If the conventional push-up type of test is used, in which a letter or other detail (Duane line, for example) is brought nearer and nearer to the eye until the subject reports the beginning of blur, obviously the point at which blur begins is not the true NPA (near point of accommodation): Because of depth of focus, this point is actually nearer than the nearest position for sharpest focus.

The *pinhole* or the stenopaic slit, when placed before the eye, increases greatly the depth of focus and keeps the blur disk on the retina small. Devices such as these make it possible to determine the visual acuity when the image normally formed by the eye is poor.

## EXERCISES

1. On the basis of the chromatic aberration of the eye (Fig. 135), would a person with uncorrected presbyopia find it easier to read at near vision in blue or in red monochromatic light?
2. The total depth of focus based upon the ability to discriminate 20/40-size letters is 2.00 diopters. What is the hyperfocal distance for letters of this size in meters for a person who cannot accommodate? In feet?
3. The words of a road sign at a distance of 25 meters are just barely legible when viewed by a person with normal vision. If the depth of focus for letters of this size in the discrimination by the average eye at this distance is 1.20 diopters, how much uncorrected myopia could a person have and still be able to read the sign at this distance?
4. What is the effective depth of field of an observer whose eyes are

1.25-diopters myopic and are uncorrected, if the depth of focus is taken to be 1.50 diopters?

5. An observer can barely resolve two close stars whose angular separation is 2.5 minutes of arc. What is his resolving power? What is the equivalent acuity in Snellen notation if tested at 20 feet?

6. A person has a visual acuity of 20/50. How high should the letter $E$ be on the chart at a distance of 20 feet in order for him to identify it? (The height of the 20/20 letter $E$ is 8.7 mm.)

## REFERENCES

1. BEDFORD, R. E., and WYSZECKI, G.: Axial chromatic aberration of the human eye. *J Opt Soc Amer*, *47:*564-565, 1957.
2. CAMPBELL, F. W.: The depth of field of the human eye. *Opt Acta (London)*, *4:*157-164, 1957.
3. HARTRIDGE, H.: The visual perception of fine detail. *Trans Roy Soc (London)*, *B232:*519-671, 1946-1947.
4. PASCAL, J. I.: Optical and visual effects of tilted lenses. In *Studies in Visual Optics*. St. Louis, Mosby, 1952, pp. 701-708.
5. SCHWARTZ, J. T., and OGLE, K. N.: The depth of focus of the eye. *Arch Ophthal (Chicago)*, *61:*578-588, 1959.
6. VOLK, DAVID: Conoid refracting surfaces and conoid lenses. *Amer J Ophthal*, *46:*86-95, 1958.

Chapter XII

# CONSIDERATIONS OF OPHTHALMIC LENSES

## 1. MAGNIFICATION PROPERTIES OF OPHTHALMIC LENSES WHEN USED WITH THE EYE

Placement of an ophthalmic lens before an eye to correct its refractive error suggests two questions related to the magnifying effect. First, how does the lens affect the otherwise normal movements of the eye in fixating objects variously located in space? Second, how does the lens affect the size of the image on the retina? Although the problems raised by these questions have much in common, the two must be dealt with separately.

Before that, however, a usable understanding of the lens should be established. With regard to angular magnification, an ophthalmic lens can be considered as if it were made in two parts (Fig. 145): first, an afocal or telescopic (Galilean) lens, whose shape

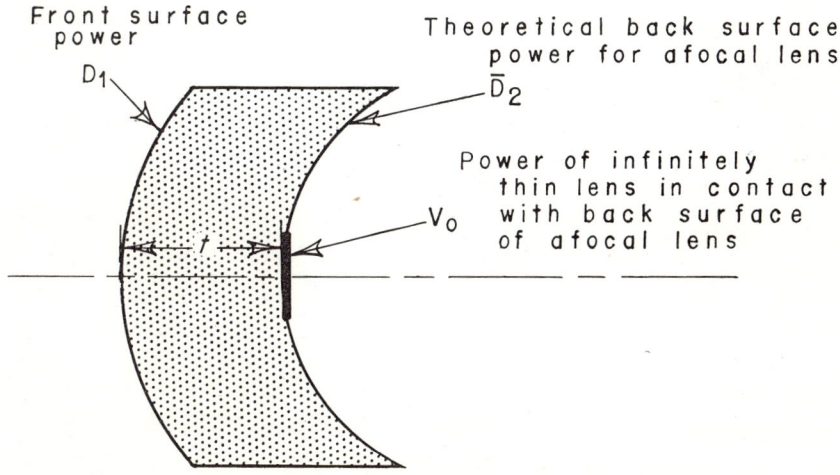

FIGURE 145. For purposes of discussing magnification, an ophthalmic lens can be considered as a combination of an afocal lens and an infinitely thin power lens.

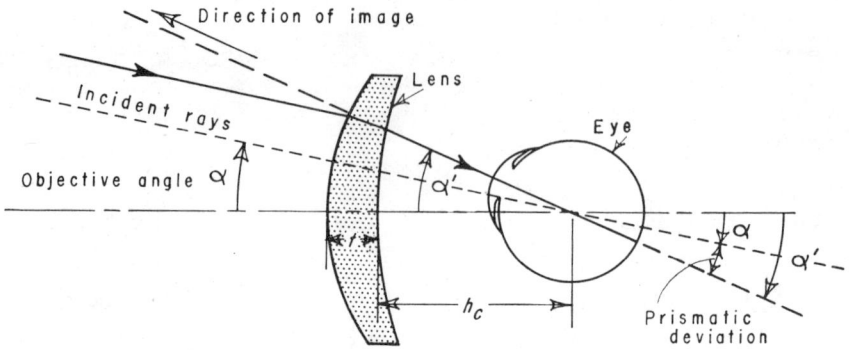

FIGURE 146. Increase in rotation of the eye necessary when it fixates the images of separated points through a converging lens.

and thickness have characteristic effects; and second, in contact with the back surface of the afocal lens, an infinitely thin lens of the exact *vertex power*, $V_o$, that is necessary to correct the refractive error.[2] Actually, the dioptric power of the back surface of the lens would be $(\overline{D}_2 + V_o)$ diopters for distant objects, in which $\overline{D}_2$ would be the theoretic power of the back surface of the afocal lens.

It has already been shown (equation 60) that the angular magnification, $A$, of an afocal lens is given by

$$A = -\frac{D_2}{D_1} = \frac{1}{1 - D_1 t/n}, \qquad \ldots (60)$$

in which $D_1$ and $D_2$ are the powers of the front and back surfaces, $t$ is the thickness, and $n$ is the index of refraction. The magnifying effect of this afocal lens is referred to as the shape factor of the total magnification, since it depends only upon the power of the front surface, $D_1$, and the thickness, $t$.

## A. Ophthalmic Lenses and Eye Movements

The angular magnification of an infinitely thin power lens, referred to the center of rotation of the eye, has already been discussed (pages 96ff). This magnification has a marked effect on the eye excursions behind the corrective lens. Let $\alpha$ (Fig. 146) be the angle through which the eye must turn in order to fixate an object point on the optical axis and then another point off axis, *without* the lens, and let $\alpha'$ be the corresponding angle through which the eye must turn behind the lens in order to fixate the images of the same two points. Then the ratio of these angles,

$\alpha'/\alpha$, is the angular magnifying effect of the lens (Fig. 76) referred to the center of rotation of the eye. The difference in the two angles, which is the *prismatic deviation* produced by the lens for any given peripheral angle $\alpha$, is then $\Delta\alpha = \alpha' - \alpha$.

→If we write $h_c$ as the distance of the imaginary thin lens (the vertex of the real lens) from the center of rotation of the eye, and $U$ as the vergence of the light incident on the corrective lens of power $F = V_o$ diopters, then the angular magnification, $A_V$, due to the thin lens, would be given as (see equation 20)

$$A_V = \frac{1 + h_c U}{1 - h_c[V_o - U]} . \qquad \ldots (72)$$

When the object points are at a considerable distance from the eye, the vergence of the incident light, $U$, is essentially zero. Then the angular magnification due to the power of the lens alone becomes

$$A_V = \frac{1}{1 - V_o h_c} . \qquad \ldots (73)$$

That part of the total magnification given by this expression is frequently referred to as the power factor. It depends only upon the vertex power, $V_o$, and the distance, $h_c$, of the lens from the center of rotation of the eye.

The total angular magnification, or the ratio of the two angles through which the eye would turn in fixating two points *with* and *without* the lens—which must take into account both the shape and power factors—would be $A = A_s A_V$, or

$$A = \frac{\alpha'}{\alpha} = \left[\frac{1}{1 - D_1 t/n}\right]\left[\frac{1}{1 - V_o h_c}\right]. \qquad \ldots (74)$$

From this formula one can calculate the eye excursion, $\alpha'$, corresponding to a given angular separation, $\alpha$, of two successive points of fixation.

*Example:* Suppose one has a +4.00-diopter lens (vertex power) at a distance of 26 mm from the center of rotation of the eye. Suppose also the front surface power of this lens is +10.00 diopters and the thickness of the lens is 4.5 mm.

The data are $D_1 = 10.00$, $t = .0045$, $n = 1.523$, $V_o = +4.00$, and $h_c = 0.026$ meter. Substituting these values into equation 74, one has

$$A = \frac{\alpha'}{\alpha} = \left[\frac{1}{1 - (10)(.0045/1.523)}\right]\left[\frac{1}{1 - (4.0)(.026)}\right],$$

or

$$\alpha'/\alpha = [1.030][1.116] = 1.149 .$$

This corresponds to a total magnifying effect of about 15 per cent. It will be seen that the shape factor of the lens contributes only 3 per cent of the total, whereas the power factor contributes 11.6 per cent. In this example, all eye excursions behind the lens are increased by 15 per cent over those which would be required if the lens were not used.

Examples of the actual eye movements for this particular lens, which are easily computed from the ratio $\alpha' = 1.149\ \alpha$, are shown in the following table.

| Peripheral angle $\alpha$ | 5° | 10° | 15° | 20° | 25° | 30° | 35° |
|---|---|---|---|---|---|---|---|
| Eye excursion $\alpha'$ | 5.7° | 11.5° | 17.2° | 23.0° | 28.7° | 34.5° | 40.2° |
| Prismatic deviation $\Delta\alpha = \alpha' - \alpha$ arc degrees | 0.7° | 1.5° | 2.2° | 3.0° | 3.7° | 4.5° | 5.2° |
| $\Delta\alpha$ in prism diopters = $1.74\Delta\alpha°$ | 1.3$^\Delta$ | 2.6$^\Delta$ | 3.9$^\Delta$ | 5.2$^\Delta$ | 6.5$^\Delta$ | 7.8$^\Delta$ | 9.1$^\Delta$ |

When the eye looks from a point on the axis of the lens to the image of a point that is 30° below the axis, the eye must turn 34.5°, or 4.5° farther than if the lens were not there. This is equivalent to a prismatic deviation of 7.8 prism diopters (prism oriented base-up).

If minus corrective lenses are used ($V_o$ is minus), then the ratio $\alpha'/\alpha$ is less than unity, consistent with the fact that all eye excursions behind the lens are smaller than those that would be required if no lens were used.

The eye movements over the entire field are uniformly enlarged or diminished, depending upon whether the corrective lens is plus or minus. The amount of increase or decrease in the eye excursions, with and without corrective lenses for any two separated points, is called a prismatic deviation. However, it is clear that the amount of prismatic deviation depends upon the angle of separation of the two points fixated. For help in remembering whether the direction of the prismatic deviation is toward or away from the visual axis, a mnemonic diagram (Fig. 147) representing converging and diverging lenses as biprisms may serve.

A rule of thumb for estimating the amount of prismatic deviation caused by an ophthalmic correction can be found easily from the concepts derived above. Certain simplifications are made: (1) the influence of shape factor is ignored; and (2) the magnifying effect of the power factor is approximated by dividing the numerator by the denominator of the ratio (equation 73), writing the result in a power series, and then neglecting higher orders of $(V_o h)$,

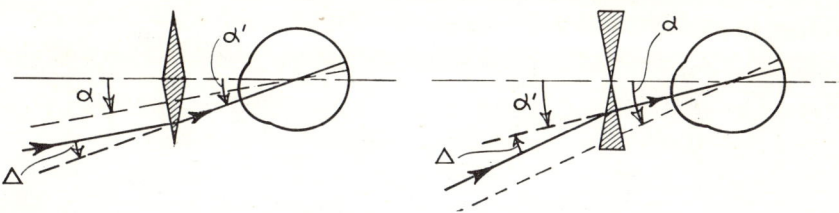

FIGURE 147. Mnemonic diagram of direction of equivalent prismatic effects of ophthalmic lenses when the eye looks down through the lens.

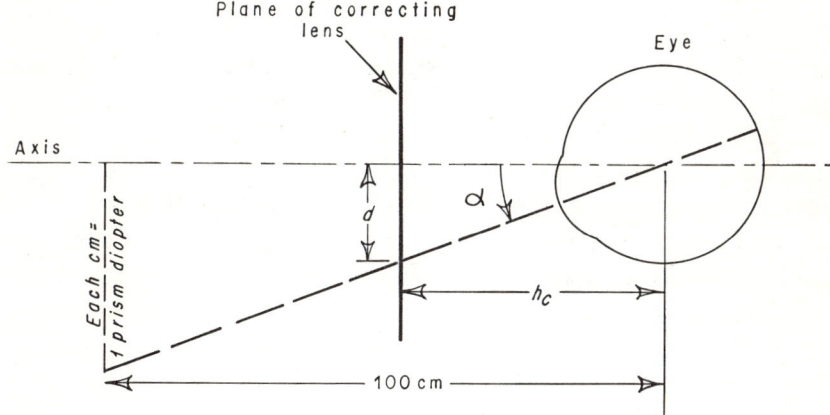

FIGURE 148. Figure used in deriving the prismatic displacement of a ray through an ophthalmic lens.

so that $\alpha'/\alpha = 1 + V_o h_c$. The prismatic deviation, then, is

$$\Delta = \alpha' - \alpha = V_o h_c \alpha. \qquad \ldots (75)$$

If $\alpha$ is given in arc degrees, the prismatic deviation is in arc degrees. If $\alpha$ is given in prism diopters, the prismatic deviation $\Delta$ is in prism diopters. Now, when the eye fixates an obliquely located point, the line of sight of the eye passes through the lens at a point a distance, $d$, from the center of the lens. One can see in Figure 148 that, from similar triangles, $\alpha$ in prism diopters $= 100\, d/h_c$. When this value is substituted for $\alpha$ in the above simplified formula, the prismatic deviation becomes $\Delta = 100\, V_o d$. In this, $d$ must be in meters. If $d$ is given in *centimeters*, then

$$\Delta = V_o d \text{ (prism diopters)}. \qquad \ldots (76)$$

Thus the prismatic deviation, in diopters, is approximately equal to the power of the corrective lens multiplied by the distance, in centimeters, from the center of the lens to that point through which the eye is looking. This is a standard formula, but it is only an approximate one. If the eye looks down through the +4.00-diopter lens, as in the preceding example, at an angle of 30°, the distance, $d$, would then be 2.6 tan 30° = (2.6) (.577) = 1.5 cm. The prismatic deviation, $\Delta$, so computed, would be 6.0 prism diopters, as against the 7.8 prism diopters found in the table. Nevertheless, for estimating the prism divergence caused by anisometropic corrections—in which there are lenses before both eyes, and in which the shape factors of the two lenses *tend* to balance one another—the above formula based on the power factor alone provides an adequate approximation. When the two eyes are lowered to the reading position (about 15°), the value of $d$ is usually estimated as about 0.8 cm. This means that the prismatic deviation caused by the lens would be about 0.8 prism diopter per diopter of lens power. Whether this resultant deviation is equivalent to a base-up or base-down prism depends upon whether the lens is plus or minus (see Fig. 147).

This relationship for the prismatic deviation (Fig. 148) produced by lenses also holds when the lens itself is displaced perpendicularly with respect to the visual axis of the eye. The displacement is equal to the distance $d$ in the formula.

## B. Ophthalmic Lenses and the Size of the Retinal Image

The problem of how the size or, more precisely, the magnification of the image on the retina is altered when an ophthalmic lens is placed before the eye to correct a refractive error is an important one in ophthalmic optics.[2] The magnification of the retinal image in the correction of a refractive error depends upon the nature of the ametropia. The ametropia may be *refractive* or *axial* in origin.

### 1. Refractive Ametropia

To emphasize the problem, consider first the extreme case of the increase in the size of the retinal image when an aphakic eye is corrected by a spectacle lens.

When the crystalline lens of an eye is removed surgically, the eye becomes hyperopic by about twelve diopters. From the point of view of simple optics, this hyperopia is the same *as if*, near the center of the crystalline lens of a normal eye, an infinitely thin lens had been inserted—a lens of such minus power that it just neutralizes the refractive power of the crystalline lens (Fig. 149).

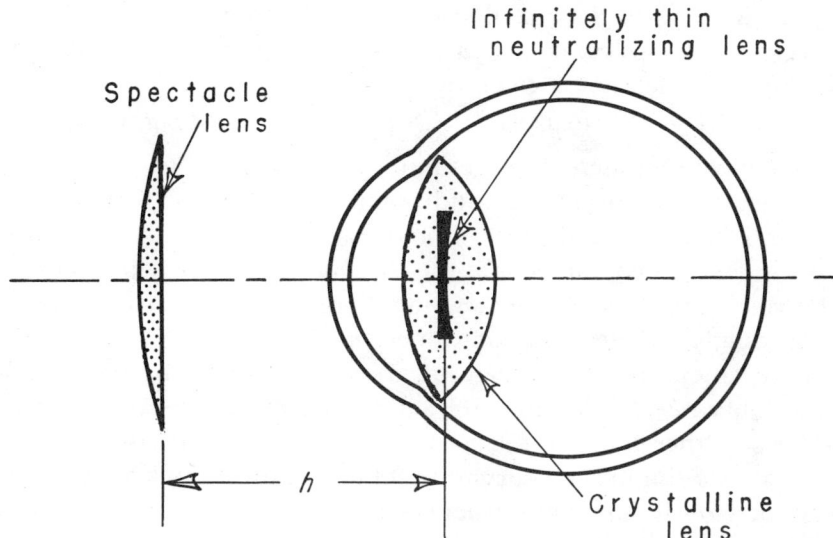

FIGURE 149. Afocal system consisting of the optical equivalent of an aphakic eye and the correcting lens. Illustration of a refractive ametropia.

This resulting hyperopia is then corrected by a spectacle lens of vertex power $V_o$, placed before the eye at a distance $h$ from the position in which the infinitely thin lens had been inserted.

Now, the combination of the correcting spectacle lens and the hypothetic infinitely thin lens constitutes an afocal system effectively in front of an otherwise normal eye; that is, the plus power of the correcting lens is such as to offset exactly the minus power of the infinitely thin lens inserted in the crystalline lens of the normal eye. For such a telescopic (afocal) system the magnification produced can be written down immediately from the results of previous discussion (on page 125, especially equation 43)—namely, for a thin correcting lens,

$$M = \frac{1}{1 - V_o h}, \qquad \ldots (77)$$

in which $V_o$, the vertex power, has been substituted for F, and $h$ has been substituted for $s$, the separation of the two components. The relationship between the powers of the two imaginary lenses of the combination is $V_o + L - (V_o L) h = 0$ (see page 123), in which $-L$ is the power of the infinitely thin lens neutralizing the power of the crystalline lens.

An average distance of the vertex of the correcting ophthalmic lens from the optical center of the crystalline lens is about 20 mm, so the magnification is given by

$$M = 1/[1 - (12)(0.020)] = 1/(1 - 0.24) = 1/0.76; \text{ or } M = 1.32,$$

which means an increase in retinal image size of about 32 per cent. For greater precision the additional magnification due to the shape factor (equation 60) should also be included. If the corrective lens is a contact lens at the corneal surface, the separation, $h$, is greatly reduced; the magnification of the retinal image is accordingly reduced to about 8 per cent.

The ametropia in the case of aphakia is called a *refractive ametropia* because the normal refractive system of the eye has been altered. Other types of refractive ametropia include astigmatism; excessive or diminished curvatures of any of the refracting surfaces; and changes in index of refraction of various media, but especially that of the crystalline lens (which is the cause of index myopia). In all these conditions the ametropia can be simulated theoretically by the concept of insertion of an infinitely thin lens somewhere in the refractive apparatus of the normal eye. This establishes a *seat* of the ametropia from which the distance to the correcting spectacle lens can be calculated.

The total magnification (for a distant object) of any ophthalmic lens of power, $V_o$, at a distance $h$ (meters) from the seat of the refractive ametropia, and with a front surface power, $D_1$, and of a thickness, $t$ (meters), is given by

$$M = \left[\frac{1}{1 - D_1 t/n}\right]\left[\frac{1}{1 - V_o h}\right]. \qquad \ldots (78)$$

This expression includes both the effect of the shape factor of the lens and the effect of the power factor and is identical to the formula (equation 74) for magnification of eye movements—except for the distance, $h$.

### 2. Axial Ametropia

Spherical ametropia also can be caused by an error in the axial length of the eye. In the normal emmetropic eye (Fig. 150), a ray from a distant obliquely located point forms an image of size $I$ on the retina. The size of this retinal image can be diagrammed as usual; in this instance an incident ray (3) is directed through

# Considerations of Ophthalmic Lenses

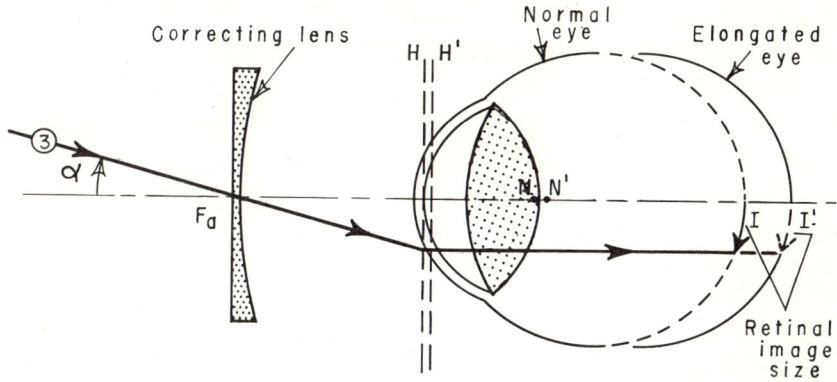

FIGURE 150. The size of the image of an axially myopic eye corrected by a lens at the anterior focal point is the same as if the eye were emmetropic (Knapp's rule).

the anterior focal point, $F_a$, of the eye, for in passing through the eye this ray must be parallel to the optic axis. The intersection of this ray and the retina determines the size of the image $I$.

If, now, this eye is elongated to cause *axial myopia* but the myopia is corrected by a thin lens placed at the anterior focal point of the eye, then the original oblique ray, still undeviated by the correcting lens, passes through the elongated eye still parallel to the axis and coinciding with its original course. So the ray intersects the retina of the elongated eye at a point which gives the retinal image, $I'$, exactly the *same size* as when the eye was normal and emmetropic.

Thus we have the rule (sometimes known as *Knapp's rule*) that when an *axially ametropic* eye is corrected by a lens placed at the anterior focal point of the eye, the magnification of the optical image on the retina is the same as that in the normal emmetropic eye. If a so-called thick lens is used, the second principal point of that lens must coincide with the anterior focal point. If the vertex of the lens is placed at this point, then the magnification of the eye is increased slightly by the effect of the shape factor. The final magnification is also slightly changed, if the lens does not coincide with this point, by the effect of the power factor, for now the distance, $h$, is measured from the anterior focal point.

One never knows precisely the position of the anterior focal point of the eye, but the approximate distance of this point is about 15 mm in front of the cornea; this is not far from where the spectacle lens is usually placed.

One does not know, either, whether the origin of a given spherical refractive error is axial, refractive, or both. Frequently, in myopia or hyperopia, the character of the optic disk as viewed with the ophthalmoscope gives a clue as to whether the spherical error is at least partially axial or refractive. The problem is complicated in eyes that have marked astigmatism in addition to a spherical error. The cylindric error can be written either as a plus or a minus cylinder, which results in a different figure for the spherical correction.

These problems of the magnification of the retinal images in correction of refractive errors have an important part in the anomaly known as *aniseikonia* (not-equal images). This anomaly is a *difference* in the magnification of the images in the *two* eyes. Because spherical ametropia may be either refractive or axial or both, one cannot estimate the aniseikonia with certainty from the amount of anisometropia present. The difference must be measured on the eikonometer. It should be clear from equation 78 that the magnification of the retinal image can be altered, within limits, by varying the surface power of the base-curve, $D_1$, by varying the thickness, $t$, of the lens, and also by varying the distance, $h$, of the lens from the eye, while keeping the corrective power, $V_o$, constant. Aniseikonic spectacle corrections are designed by selecting different base-curves, thicknesses, and distances of lenses from the eyes, in order to equalize the magnification of the images in the two eyes.

## 2. AFOCAL LENSES

The difference in the magnification of the images in the two eyes can be corrected by the placing of suitable afocal magnifying lenses (sometimes called *iseikonic lenses*) over regular spectacle lenses. These afocal lenses, when placed before an eye, change the magnification of the retinal image without introducing any change in power of the refractive correction. The angular magnification of the afocal lens (page 148) is given by the shape factor alone, since the vertex power, $V_o$, is zero:

$$A = 1/(1 - D_1 t/n). \qquad \ldots (79)$$

However, at the same time the surface powers of this afocal lens are related by

$$D_1 + D_2 - D_1 D_2 t/n = 0, \quad \ldots (80)$$

where, as before, $D_1$ is the power (diopters) of the front surface of the lens, $t$ is the thickness in meters, and $n$ is the index of refraction. For distant objects the magnifying effect of such lenses is independent of their distances from the eye. The effect is conveniently expressed in percentage of magnification, in which $m\% = 100 (M - 1)$, or approximately, $m\% = D_1 t/n$, if $t$ is expressed in centimeters. The following table gives typical values for the percentage magnifications for afocal lenses having given powers of front surface, $D_1$, and given thicknesses, $t$.

PERCENTAGE MAGNIFICATION OF SELECTED AFOCAL LENSES

| Thickness, mm | Front base-curve, diopters | | | | | | |
|---|---|---|---|---|---|---|---|
| | 3.00 | 4.50 | 6.00 | 7.50 | 9.00 | 10.50 | 12.00 |
| 1.0 | 0.20 | 0.30 | 0.40 | 0.50 | 0.60 | 0.69 | 0.78 |
| 2.0 | 0.40 | 0.60 | 0.80 | 0.86 | 1.20 | 1.40 | 1.60 |
| 3.0 | 0.60 | 0.90 | 1.20 | 1.50 | 1.80 | 1.91 | 2.42 |
| 4.0 | 0.80 | 1.19 | 1.61 | 2.01 | 2.43 | 2.84 | 3.26 |
| 5.0 | 1.00 | 1.50 | 2.01 | 2.52 | 3.00 | 3.56 | 4.10 |

When these afocal lenses are made with spherical surfaces, they magnify uniformly in all meridians and are frequently referred to as "overall" afocal magnifying (size) lenses. Meridional afocal magnifying lenses, which magnify in one meridian only and not in the meridian at right angles, can be obtained by making both surfaces cylindric with parallel axes. This type is called bitoric. The axes of the two surfaces must be precisely aligned; otherwise, unwanted astigmatism will result.

A meridional afocal lens causes a small rotary deviation of the images of all lines not parallel or perpendicular to the axis of the lens. Such a lens, placed at an oblique axis before the eye, causes a rotary deviation of the images of lines which are vertical and horizontal toward the meridian of magnification (see Fig. 133). It can be shown[2] that the magnitude of the rotary deviation of the image of a vertical line from the true vertical, $\delta_v$, for a meridional lens of $m$ per cent magnification placed at an axis, $\phi$, is given, in arc degrees, by $\delta_v = 0.29 \, m \sin 2\phi$. An equal but opposite deviation affects images of horizontal lines. The greatest deviation occurs when the axis of the lens is at the oblique angles 45° and 135°. At these angles the rotary deviation is about 0.3° per 1 per cent of magnification. The rotary deviation of these images is important in ophthalmic optics, insofar as it relates to the effects of astigmatism at oblique axes, the distortion of stereoscopic perception of space, cyclophorias, and cyclofusional movements of the eyes.

## 3. TRIAL-CASE LENS DESIGN

In the modern trial case of precision test lenses the principle is maintained that the total power of the combination of any spherical lens with any cylindric lens is specified as the vertex power from the back surface of the lens nearest the eye. It can be shown algebraically that for the power of the combination to be so specified, certain requirements are necessary in the design of the trial-case lenses and in the manner in which these lenses are placed before the eye.

To begin with, both spheres and cylinders must always be placed in the same relation to each other. Usually the sphere is placed nearer to the eye and the cylinder in front of it (away from the eye). Also, the sphere should always be placed so that its vertex is the same distance from the cornea. Most trial-lens frames are provided with a corneal-aligning device so that this requirement can be met. The intended distance from lens to cornea varies somewhat among different optical manufacturers, but it is in the range from 12 to 14 mm. Second, the powers of front surfaces and the thicknesses of all spheres must be the same. Only the back curve is changed to provide the specified vertex spherical power. Essentially this makes the shape factor in the formula for vertex power and for magnification the same for all the spheres. Each cylinder must be so mounted in its ring that, when it is placed in front of any sphere, the separation of the two inner surfaces is constant. Any cylindric lens, placed at the specified distance before any spherical lens, then gives an astigmatic correction in *vertex dioptric power relative to the vertex of the spherical lens*. Thus, though the actual power of each cylinder may not be precisely that marked on the handle, when it is used with any sphere its *effective power* is as if its vertex were at the vertex of the sphere. With this design the trial-case lenses provide precision optical systems. Even when the prescription calls for no spherical correction, the "zero" diopter sphere, found in the trial set, should be used with the cylinder lens for utmost precision—except that for low powers this precision is perhaps unnecessary, especially if a tolerance of ±0.25 diopter is to be allowed in refracting the eye.

## 4. CORRECTED-CURVE OPHTHALMIC LENSES

When the eye turns to observe an obliquely situated object through an ophthalmic lens, the rays from that object which pass through the pupil of the eye strike the lens at an oblique angle, and the image produced by the lens tends to deteriorate because of the several aberrations discussed previously.

Figures 151 and 152 illustrate a hyperopic eye and a myopic eye, respectively, corrected by spectacle lenses. When the eye is looking through the center of the lens, the focal point, $F$, is also the point conjugate to the retina of the eye—that is, the far point. When the eye turns to observe an obliquely located object, the point conjugate to the retina should lie on the far-point sphere with radius $CF$. If the refractive error of the eye is to remain fully corrected, the focal length of the lens for this oblique angle should also fall on this spherical surface. Usually the image formed by the lens does not fall on this surface. Also, the image is astigmatic, so tangential and radial astigmatic focal lines fall on other surfaces.

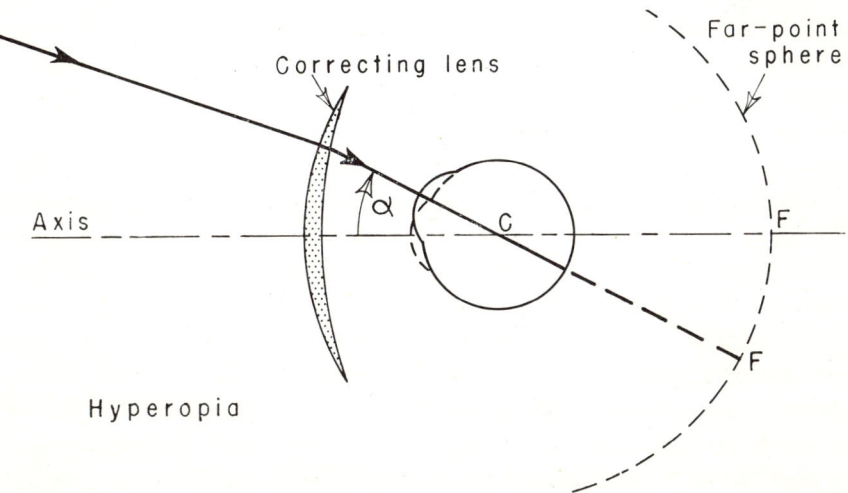

FIGURE 151. Ideal far-point sphere of a corrected hyperopic eye associated with excursions of the eye.

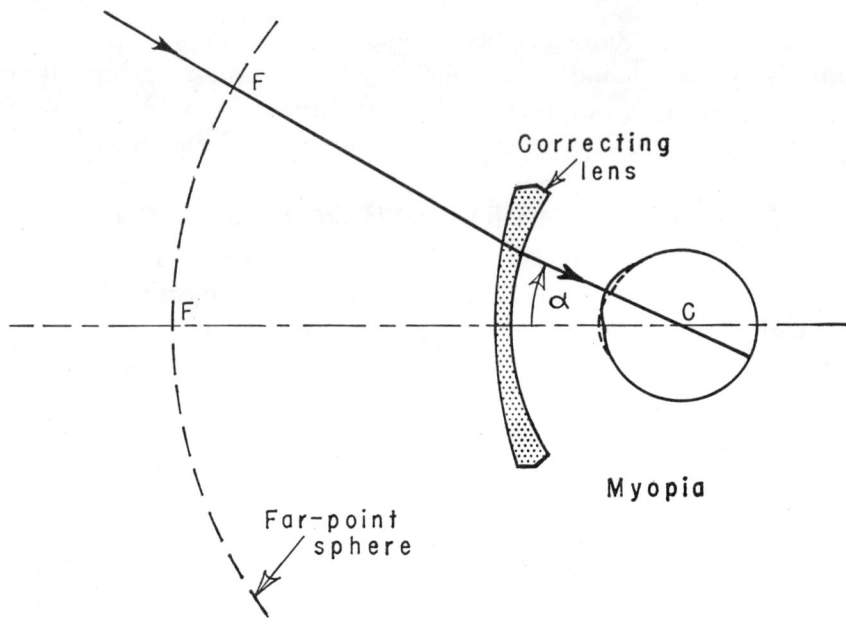

FIGURE 152. Ideal far-point sphere of a corrected myopic eye associated with excursions of the eye.

By selecting a particular base-curve of the correcting lens, depending upon the power, it is possible to minimize and in some instances to eliminate the astigmatic error and spherical error for a specified oblique angle. If the radial and tangential astigmatic focal lines can be made to coincide, the images become stigmatic, even though they may not lie quite on the far-point sphere. Lenses designed to approximate this condition are called by various trade names such as Punktal,® Orthogon,® and so forth.

Tscherning is credited for having worked out the approximate theoretic solution to the problem of obtaining lenses that eliminate the astigmatism of oblique incidence. He had to make certain assumptions: that the lens has zero thickness, that the object is at infinity, that the oblique angle is small, and that the vertex distance from the lens to the center of rotation is 25 mm. The curve showing the relationship between the power of the front surface of the lens and the power of the lens itself for which the astigmatism was eliminated is reproduced in Figure 153. It is known as the Tscherning ellipse. That part of the curve utilizing higher powers of the front surface and known as the Wollaston branch is little

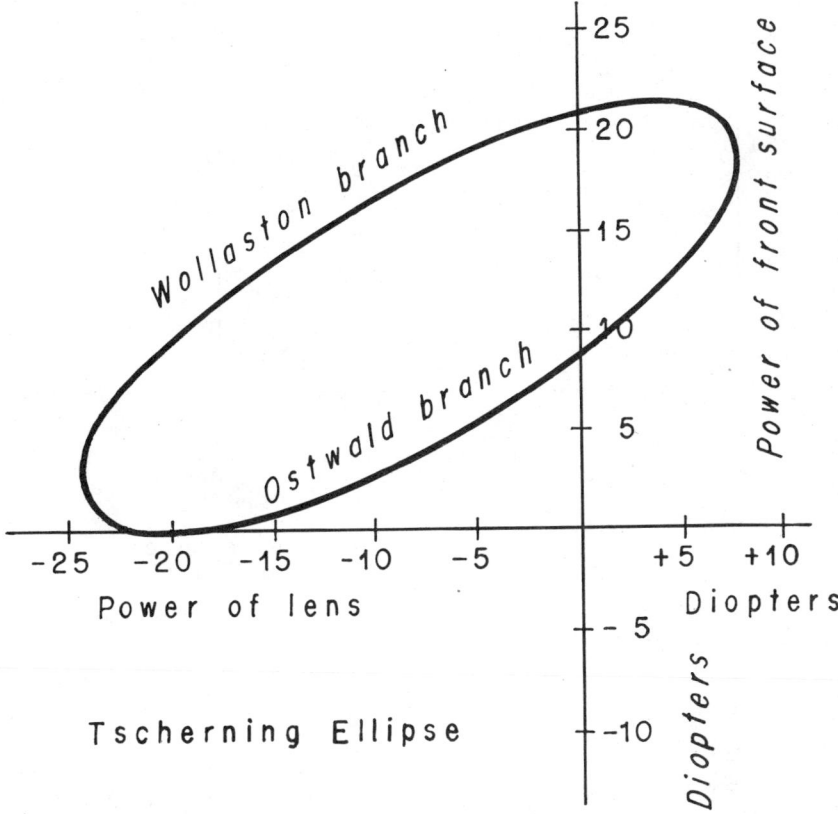

FIGURE 153. Tscherning ellipse, showing approximate base-curve required if an ophthalmic lens is to have no astigmatism for rays at oblique incidence.

used because the base-curves are too high. That part of the ellipse which utilizes the lower powers of the front surface is known as the Ostwald branch and has served as a basis for much ophthalmic lens design. Note that for plus lenses the base-curve must be increased as the power of the lens is increased. As the power of a minus lens increases, the power of its front surface must be decreased; that is, the surface must become flatter. Theoretically, each spectacle lens, depending upon its power, should have a different base-curve.

An enormous amount of labor has been put into calculation of the power errors produced by real lenses (that is, those having thicknesses) for particular oblique angles. Typical results of cal-

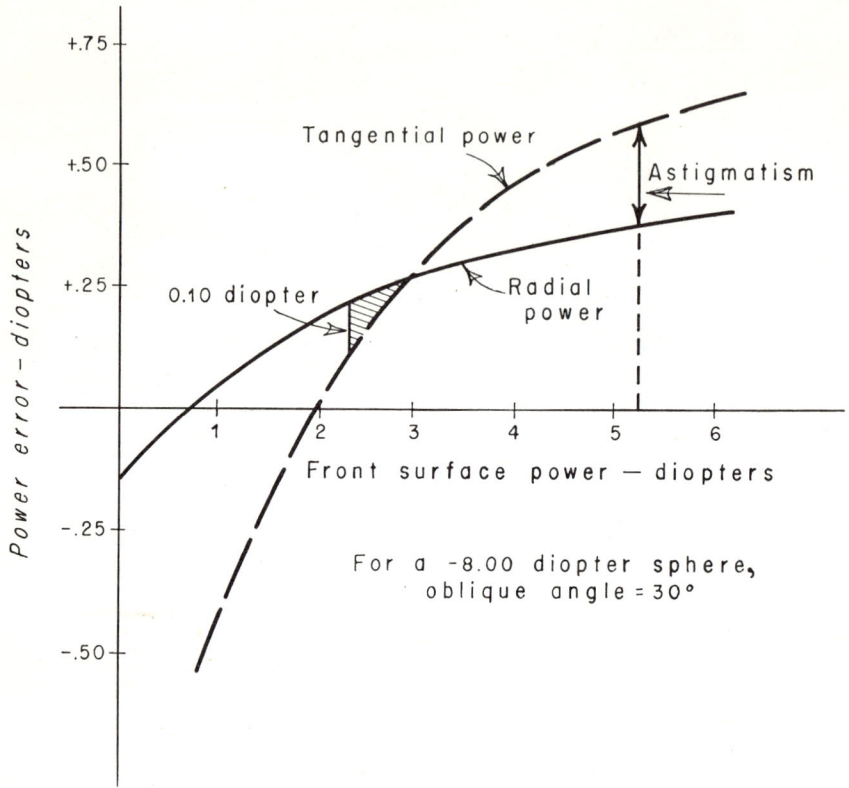

FIGURE 154. Relationship between the power errors of a −8.00-diopter sphere for different powers of its front surface and for an oblique ray at a 30° angle.

culation of the power errors (Tillyer) for a lens of −8.00 diopters and an oblique angle of 30° are reproduced in Figure 154. Note that when the base-curve is about 3.00 diopters, the astigmatism is eliminated but a spherical power error of +0.25 diopter persists. It might be proposed to minimize both the spherical and astigmatic errors by allowing an astigmatic error of 0.10 diopter. Accordingly, the base-curve would be reduced to about 2.25 diopters, which allows an astigmatic error of 0.10 diopter but reduces the spherical equivalent error to 0.15 diopter. Calculations for curves of this kind have been made for a large number of possible lenses (including astigmatic corrections), and from these the optimal base-curves have been selected.

Figure 155 illustrates the results of other recent calculations.

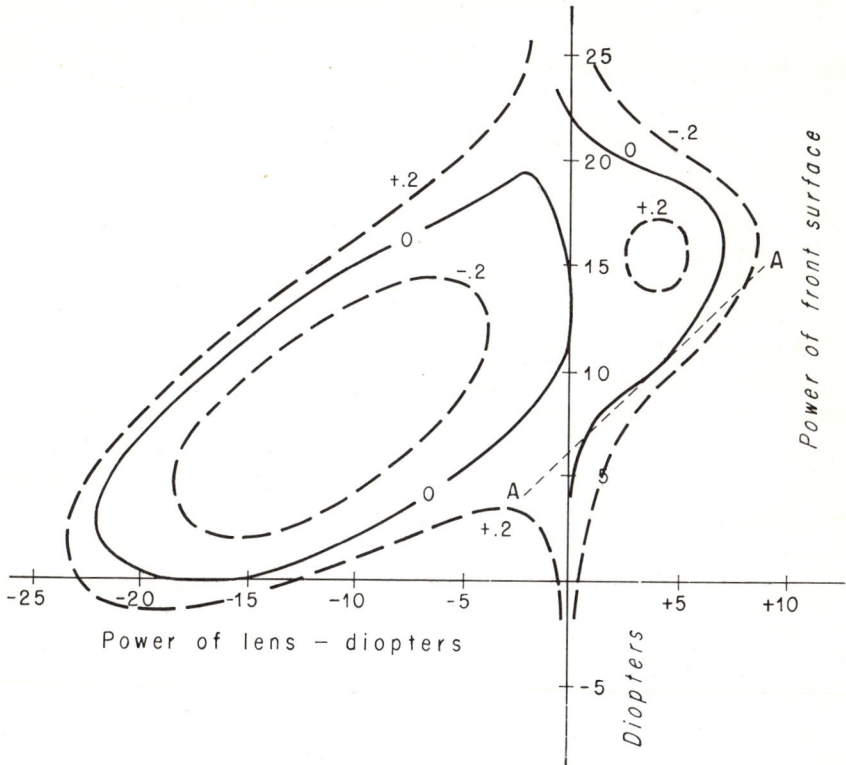

FIGURE 155. Calculated relationship (lines $O$) between different lens powers and the front-surface powers required to eliminate astigmatism for rays of oblique incidence (with added curves for tolerance of ±0.2 diopter).

In these the lenses were given realistic dimensions: namely, thickness of 1 mm for minus lenses and $(1 + 0.72\ V_o)$ mm for plus lenses; the distance of the lens from the center of rotation of the eye was stated as $(25 - 0.2V_o)$ mm; and the oblique angle was specified as 30° and the calculations made for a distant object. The solid-line curves indicate the relationship of the power of the front surface curve, $(D_1)$, of a lens to a given vertex power, $V_o$. The interrupted lines are the plus and minus 0.2-diopter astigmatic tolerance curves. It was postulated on the basis of these calculations that all plus lenses and minus lenses to about $-2.50$ diopters could be designed with a back-surface power $(D_2)$ of $-6.00$ diopters, for then the astigmatism would be well below the 0.2-diopter tolerance. Such a series would follow the line $A$—$A$ on the graph. For a near visual distance of 30 cm the ordinates should be dropped 2.00 diopters.

The manufacture of lenses with individual base-curves for each power would be costly and impractical, so lenses are usually placed in groups according to their power. The lenses in each group have the same base-curve and are limited to a range of spherical and astigmatic powers. Only those lenses in the middle of the group are the more truly corrected-curve lenses.

It should be clear from this discussion that, while these corrected-curve lenses are highly desirable, the proper curve varies somewhat with object distance and with the vertex distance of the lens from the eye. In spite of errors that may occur, today the ophthalmic lenses of finest quality are the corrected-curve lenses. However, an experimental evaluation of the superiority of corrected-curve lenses as compared to noncorrected-curve lenses has never been made. Except for lenses used to correct aphakia, many persons cannot discriminate the superior qualities of the corrected-curve lenses. In line with this discussion, however, is the fact that myopes are very sensitive to changes in base-curve, and the majority prefer as flat a curve as possible. The reason for this is by no means clear.

Of course, aniseikonic spectacles, in which differences in the base-curves and thicknesses of the two lenses for the two eyes are intended to correct the difference of magnification, ignore the scheme of corrected-curve lenses.

The optical problems in the design and use of bifocal lens segments will not be discussed in this monograph; the subject is a large one, and the reader should consult texts on refraction. In this connection, mention should be made, however, of the omnifocal lens[4] and the varilux lens,[1] in which the plus power increases continuously from the upper portion of the lens to the lower portion. Such a lens permits the early and late presbyope to have a continuous "add" for all distances from far to near vision, but it has rather bad optical quality off the axis of the lens.

## 5. CONTACT LENSES

The types of contact lenses available today are the results of an evolutionary development, so much of the optical theory of contact lenses to be found in older textbooks is no longer wholly pertinent. Contact lenses are important for several reasons beyond cosmetic considerations: (1) they eliminate the irregularities of the cornea in such conditions as keratoconus and high astigmatic errors; (2) they eliminate the need for heavy spectacles in high refractive errors; (3) they permit a wide visual field unrestricted by the size of spectacle lenses; and (4) they eliminate the peripheral aberrations inherent in spectacle lenses. In aphakia, contact lenses greatly reduce the magnification of the image in the retina of the eye.

The difference between the indices of refraction of the plastic material of which the contact lens is made (and the thin layer of tears between this lens and the cornea) and of the cornea is very small, so the refractive power of the cornea is greatly reduced, if not almost eliminated. The front surface of the contact lens becomes, in effect, the new corneal surface. The radius of the back surface of the lens is usually fixed by the corneal curvature, so the power of the lens must be obtained through proper selection of the curvature of the front surface of the contact lens. Consequently, the refractive power of this front surface must include the normal power of the cornea plus the correction for a refractive error. In many instances the determination of the correction must be made with a trial contact lens in place, since the corneal astigmatism has been eliminated by the contact lens. The actual dioptric value of this correction is different, of course, from that determined by trial-case lenses, since now the correction is so much nearer the eye. The power is increased for hyperopic correction and decreased for myopic correction.

As pointed out earlier (page 203), when a contact lens is substituted for a spectacle lens to correct a given refractive error, a *change* in the magnification of the retinal image results. The direction of the change depends upon whether the spherical error is a hyperopia or a myopia. If the eye is hyperopic, substitution of the contact lens results in *decreased* magnification (the extreme case being that in the correction of aphakia[3]); if myopic, magnification is *increased*.

The actual remaining magnification or size of the retinal image compared to that of the emmetropic eye depends, of course, upon whether the ametropia is refractive or axial. In refractive ametropia, if the eye is hyperopic, even though the magnification is reduced by the use of the contact lens there still remains a small magnification of the retinal image (especially in aphakia). The converse is true of refractive myopia. On the other hand, in axial ametropia the retinal image is smaller in hyperopia and larger in myopia when the contact lens is used. This follows because the corrective lens (the contact lens) is displaced toward the eye from the anterior focal point, at which location a corrective lens would produce an image on the retina the same size as it would be if the eye were emmetropic (Knapp's rule).

In considering the magnifying effect of contact lenses, the shape factor must also be taken into account. Even though the thickness is small, the power of the front curve is so high that the magnifying effect may amount to 2 to 4 per cent, depending upon the particular dimensions of the lens.

## EXERCISES

1. A +6.00-diopter thin spherical lens is placed before the eye 25 mm from the center of rotation. What is the angular magnification of all eye excursions behind the lens? If the eye fixates a point 30° to the right of the optic axis of the lens, through what angle has the eye turned? What is the angular prismatic deviation in degrees? In prism diopters?

2. For the lens in the problem above substitute one of −6.00 diopters. Answer the same questions.

3. Suppose the +6.00-diopter lens is placed before the right eye and the −6.00-diopter lens before the left eye. What is the total prismatic deviation between the eyes for the 30° angle?

4. Using the approximate formula (76) for prismatic deviation, calculate the prismatic deviations in the above three examples. What point is being illustrated?

5. If in problem 3 the eyes were lowered to fixate a point 30° below the optic axes of the two lenses, what is the vertical divergence introduced, rigorously and approximately?

6. Astigmatism is a refractive ametropia. If the seat of an astigmatism is entirely at the cornea and is corrected by a +4.00-diopter thin cylinder lens at 12 mm from the cornea, what is the meridional magnification of the retinal image?

7. If the cylinder lens in the previous example were placed before the right eye at axis 45°, what would be the approximate rotary deviation

of the images of vertical lines? Would the direction of rotation be excyclotorsional or incyclotorsional?

8. A person wears a +5.00-diopter lens before each eye, at 25 mm from the centers of rotation. These lenses are centered for distant vision (visual axes of eyes parallel). The person then looks at print at 40 cm. What is the total prismatic deviation introduced by the lenses? Assume the interocular separation is 64 mm. How much would each lens have to be decentered so that there would be no prismatic deviation at this reading distance? Hint: Determine displacement of lines of sight by similar triangles.

## REFERENCES

1. MAITENEZ, B.: Four steps that led to the varilux. *Amer J Optom, 43:*441-450, 1966.
2. OGLE, K. N.: *Researches in Binocular Vision.* Philadelphia, Saunders, 1950, pp. 122ff. Reprinted 1964, New York, Hafner.
3. OGLE, K. N., BURIAN, H. M., and BANNON, R. E.: On the correction of unilateral aphakia with contact lenses. *Arch Ophthal (Chicago), 59:* 639-652, 1958.
4. VOLK, D., and WEINBERG, J. W.: The omnifocal lens for presbyopia. *Arch Ophthal (Chicago), 68:*776-784, 1962.

*Chapter XIII*

# PRINCIPLES OF CERTAIN OPHTHALMIC DEVICES

SEVERAL special types of ophthalmic instruments deserve separate consideration. These depend on the basic optical principles discussed in the preceding pages, and in many instances their applications of the principle involved are quite simple. The details of the instruments may be found in other sources and will not be discussed here, since this monograph is for the purpose of familiarizing the reader only with the basic principles.

## 1. OPHTHALMOSCOPY

Normally, as we look at another's eye, the pupil opening appears black. No area of the retina can be seen, even though it may be highly illuminated. The reason for this is that those particular rays of light reflected from the retina which would enter the examiner's eye must follow approximately the same paths as those rays which enter the eye to illuminate the retina. These particular incident rays are usually prevented from entering the eye by the head of the examiner. To overcome this, von Helmholtz[3] used a plane-parallel glass to reflect light from a light source (Fig. 156) into the eye, approximately along the axis joining the pupils of the subject's and examiner's eyes. The rays then reflected from the subject's retina passed through the glass plate to the examiner's eye, thus permitting the retina to be seen. With the modern ophthalmoscope the examiner views the subject's eye through a small hole in a mirror or over the edge of a total reflecting prism which reflects light into the subject's eye.

In *direct* ophthalmoscopy, as was mentioned in Chapter VI, the examiner effectively uses the dioptric system of the subject's eye as a type of hand magnifier to view the retina. If the subject's eye is emmetropic, then the emerging rays reflected from the retina

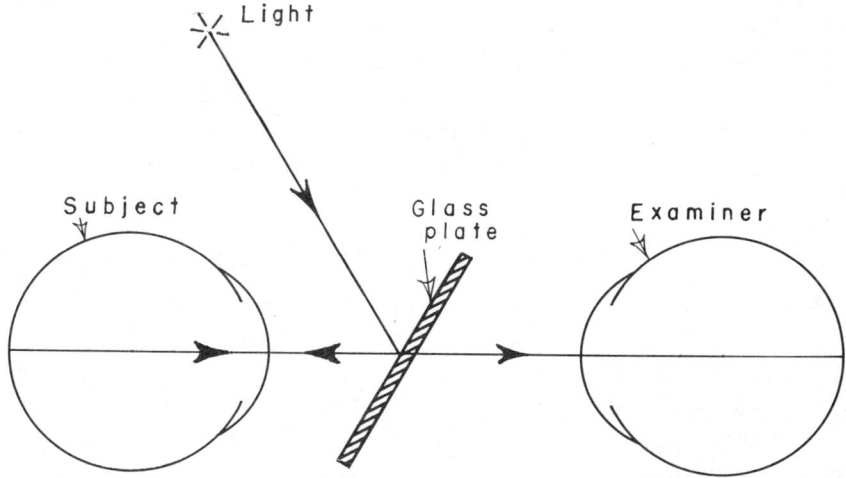

FIGURE 156. Principle of direct ophthalmoscopy.

are parallel, and the image of the retina as seen by the examiner is at infinity. Small degrees of ametropia are corrected by the insertion of compensating lenses before the examiner's eye. The advantage of the direct method is its simplicity. The objection is that only a small portion of the retina can be seen at one time. The magnification is greatly increased in high myopia.

In *indirect* ophthalmoscopy, the parallel (or nearly parallel) rays emerging from the eye are focused by a large field lens of about +13 diopters held before the eye (Fig. 157). Thus a real image of the fundus is formed, nearly at the focal plane of the field

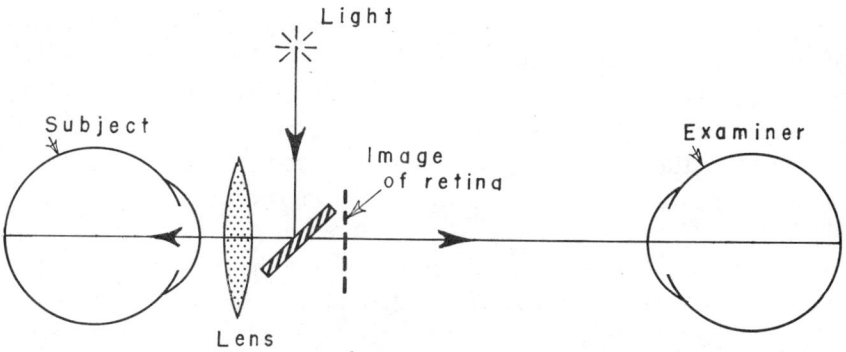

FIGURE 157. Principle of indirect ophthalmoscopy.

lens; the examiner observes this inverted image of the retina. To do so, the examiner must be sufficiently far from the lens (about 60 cm) or must use a low-power type of ocular. The advantage of this method is the increased size of field or area of the retina visible. The disadvantage is the lower magnification.

## 2. RETINOSCOPY (SKIAMETRY)

Retinoscopy is a relatively simple and objective method of estimating the refractive error of an eye.[7] The retinoscope (which is essentially a refinement of the ophthalmoscope for indirect ophthalmoscopy) is used to project a beam of light into the subject's eye so that a small area on the retina is highly illuminated. The light reflected from this spot on the retina is then observed by the examiner through a small aperture in the retinoscope—a small hole in the mirror which projects the light into the subject's eye. As the light from the retinoscope is moved back and forth across the pupil of the subject's eye, the examiner observes the apparent motion of the light in the pupil as lenses of different powers are placed before the eye. The critical point of the test occurs when that particular lens is found for which the apparent motion of the light reflex from the spot reverses its direction. For this light vergence, the conjugate point of the illuminated spot on the subject's retina coincides with the pupil of the *examiner's eye*.

It is often difficult for the student in ophthalmoscopy to understand the theory of the retinoscope. In part the difficulty has been due to the fact that the change in motion of the reflex has been sought in events occurring in the subject's eye, whereas the secret of the reversal of the movement of the retinal light reflex lies in the image formation by the emerging light *at* the examiner's eye.

When the standard self-luminous retinoscope is turned in such a way that the light beam passes downward across the subject's pupil, as shown in Figure 158, the light spot on the subject's retina also moves down. This direction of movement of the light spot on the retina follows because the light source from the retinoscope effectively moves upward. This direction of movement is the same in all eyes, *regardless of the refractive error*. In the following discussion it is convenient to assume that a light spot on the retina of the

*Principles of Certain Ophthalmic Devices*

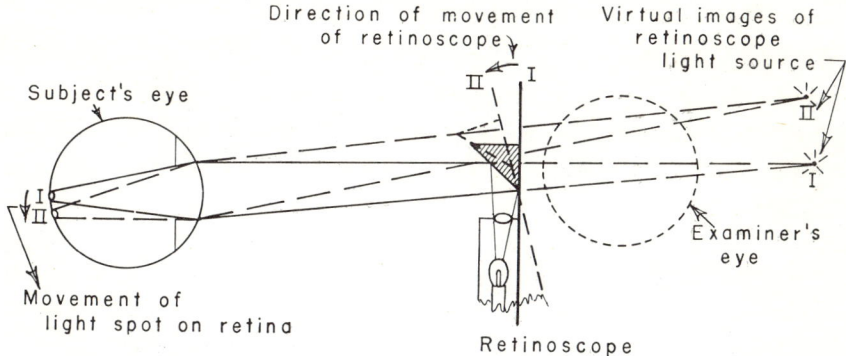

FIGURE 158. Diagram illustrating movement of the light spot on the retina associated with movement of the self-illuminated retinoscope.

subject's eye moves *downward*. The discussion concerns the position of the image of the light originating from the illuminated area on the retina, as formed by the dioptric system of the eye.

The vergence of the light emerging from an illuminated retinal spot of an emmetropic unaccommodated eye is *zero*; that is, the rays emerging from the eye are parallel. To focus this emerging light to the examiner's eye (Fig. 159), we must place before the subject's eye a lens whose focal length is equal to the distance from the lens to the examiner's eye. If the examiner maintains a constant distance of, say, 66.6 cm, the power of this spherical lens must be +1.50 diopters. In practice, a +1.50-diopter lens is placed at the outset before *any* eye to be tested, and the examiner maintains his distance from this lens.

Under these circumstances, if the subject's eye is *myopic*, the

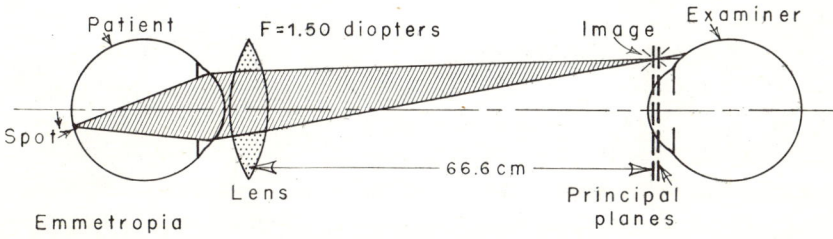

FIGURE 159. Relationship between subject's retina and the reflected image at examiner's eye for the point of reversal in retinoscopy.

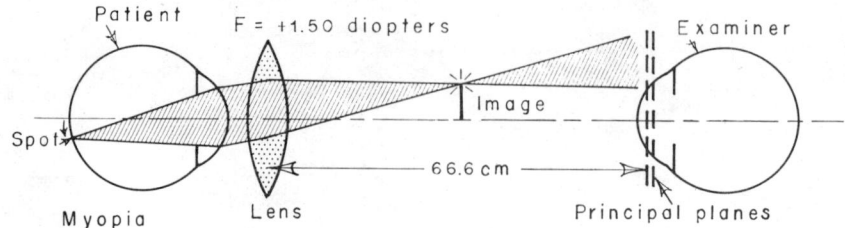

FIGURE 160. The image of light reflected by the retina of a myopic eye lies in front of the examiner's eye in retinoscopy.

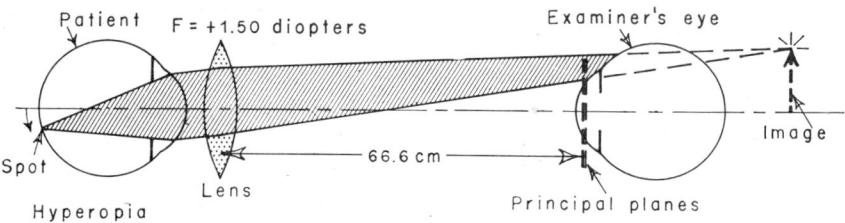

FIGURE 161. The image of light reflected by the retina of a hyperopic eye lies behind the examiner's eye in retinoscopy.

image of the illuminated retinal spot would lie in front of the examiner's eye, as shown in Figure 160.

If the subject's eye is *hyperopic*, then the image of the illuminated retinal spot would lie behind the examiner's eye, as shown in Figure 161.

The direction of the movement of this image is always upward, opposite to the downward movement of the light spot on the subject's retina. This direction of movement is the same, regardless of whether the subject's eye is myopic or hyperopic. To neutralize the effect of the myopic or hyperopic refractive error, particular ophthalmic lenses are then added before the +1.50-diopter sphere until the image of the illuminated retinal spot coincides with the pupil of the examiner's eye. The refractive error equals the power of the lenses used in conjunction with the +1.50-diopter sphere.

To explain the reversal of the motion of the blurred light image in the examiner's eye at this critical point, we must review the basic optics involved in image formation by converging lenses. If the patient's eye is *myopic* (Fig. 160), then the point conjugate to that retina lies in front of the examiner's eye. The rays from this

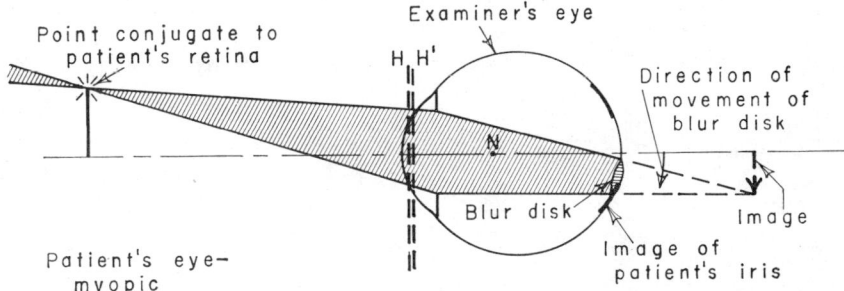

FIGURE 162. Direction of movement of the image blur spot in examiner's eye for light reflected from the retina of a myopic eye.

image which diverge into the examiner's eye are illustrated schematically in Figure 162.* Since the accommodative state of this eye is constant, the image formed by the dioptric system of the examiner's eye is behind the examiner's retina, but the rays are intercepted, resulting in a blur disk on the retina. The movement of this blur disk on the examiner's retina will be *downward*, but the examiner *perceives* the light blur as moving *upward* across the image of the subject's pupil, therefore in the direction opposite to, or "against," the movement of the light beam from the retinoscope across the patient's pupil.

If the patient's eye is *hyperopic* (Fig. 161), then the point conjugate to that retina lies behind the examiner's eye. The rays from this image, which converge into the examiner's eye (and therefore the equivalent object distance, $u$, in the usual formula for object and image distances, is minus), are illustrated schematically in Figure 163. Here the image formed by the dioptric system of the examiner's eye lies in front of his retina, but again the rays are intercepted by the retina, resulting again in a blur disk. The movement of this blur disk, however, is *upward*, but the examiner *perceives* the light as moving *downward* across the image of the patient's pupil, therefore in the same direction as, or "with," the movement of the light beam from the retinoscope across the patient's pupil.

If the patient's eye is emmetropic, or the ametropia has been

---

*These diagrams (Figs. 162, 163, and 164) are greatly exaggerated in order to make clear the principles involved. The actual cone of light from the patient's pupil will obviously be very small.

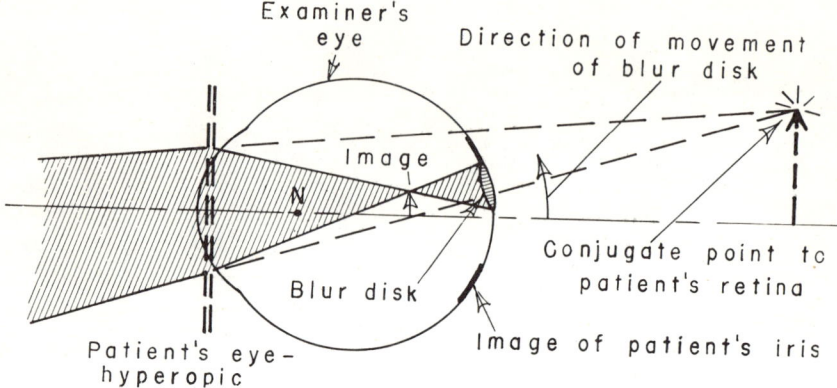

FIGURE 163. Direction of movement of the image blur spot in examiner's eye for light reflected from the retina of an unaccommodated hyperopic eye.

fully corrected by suitable trial lenses, then the image of the illuminated retinal spot coincides with the pupil of the examiner's eye (Fig. 164). As the image of this spot passes the pupil of the examiner's eye, the examiner sees a flash of light in the patient's pupil—but without apparent movement, for the blur of light on the examiner's retina is now very large and moves very fast. This event occurs at the point of image reversal and marks the presence of that condition which is sought in determining the refractive error of the eye with the retinoscope.

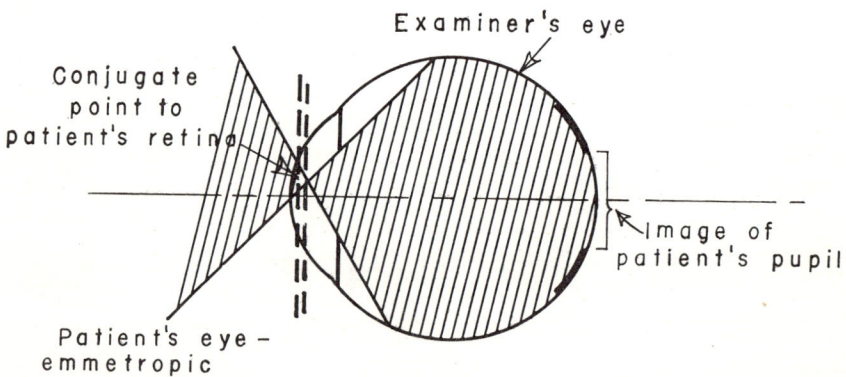

FIGURE 164. Point of reversal in retinoscopy—the flooding of light in examiner's eye when the image of light reflected from the retina of the subject's eye lies at the examiner's pupil.

If the patient's eye (with the +1.50-diopter lens) is myopic, the examiner sees what is termed an "against movement" of the retinal light reflex across the pupil of the eye. The examiner then adds lenses of increasing minus power until the reversal point is reached. Conversely, if the eye is hyperopic, the examiner sees a "with movement" of the retinal light reflex across the pupil of the eye and so adds lenses of increasing plus power until the reversal point is reached.

The optical theory of retinoscopy as given above is entirely correct. In practice, however, certain aspects of the retinal reflex should be noted: The examiner observes closely the pupil of the subject's eye, but in doing so he cannot focus on the light spot on the retina or any portion of the retina as in ophthalmoscopy because the image of the reflected light from the retinal spot is too near the examiner's eye; hence the image on his retina is much out of focus. If the light beam from the retinoscope is on the axis between the subject's and examiner's pupils, the examiner sees only a pupil filled with a bright reddish light. As the retinoscope beam is moved across the subject's eye, the intensity of the light in the pupil decreases and darkness or a dark shadow follows rapidly. When the reflected light beam from the subject's eye no longer falls within the examiner's pupil, the subject's pupil appears black. The rate of movement of the light reflex and following shadow decreases as the final image from the spot gets farther from the examiner's eye. The maximum rate of movement occurs when that image coincides with the examiner's pupil—that is, when the neutral point has been obtained by suitable choice of correcting lenses in front of the +1.50-diopter field lens.

Although mirror retinoscopes that are not self-illuminated are rarely used today, one can easily show that a *plane*-mirror retinoscope reflecting the light from a lamp near the patient into the patient's eye functions as the self-illuminating retinoscope does. Also a retinoscope with a *concave* mirror, if its focal length is relatively short, can be used to reflect light into the eye, but with this retinoscope the retinal reflex for the myopic and hyperopic eyes moves in directions *opposite* to those seen with the plane-mirror and self-illuminating retinoscopes.

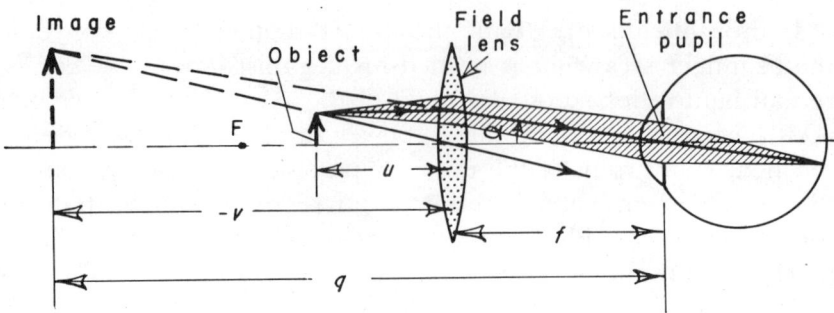

FIGURE 165. Badal principle, whereby the angular size of the image and the luminance of the object remain constant, irrespective of its axial position.

## 3. BADAL PRINCIPLE

When a card of print is moved toward the eye—as, for example, in determining the nearest point at which print of a given size is still legible—the angular size of the letters increases, changing from what it was at the normal reading distance. Following a principle pointed out by Badal,[1] this change in angular size can be eliminated by use of a *converging* field lens placed before the eye (Fig. 165) at such a distance that the posterior *focal* point of the lens coincides with the entrance pupil of the eye. As the test card or object is moved along the axis of the lens, its image moves over a wide range of distances from the eye. When the object is precisely at the anterior focal point of the lens, $F$, the image is at infinity. In diagramming the rays in the figure, that particular ray from the object which is incident to the lens *parallel* to the optic axis proceeds to the center of the entrance pupil, since the principal focal point of the lens is placed to coincide with the entrance pupil of the eye. This ray, then, follows the same course, irrespective of the position of the object on the axis. Hence the angular size of the image of the object remains the same, irrespective of the position of the test target.

If $q$ is the distance of the image from the eye, then $Q = 1/q$ is the vergence of the light from the image at the eye (or to a reference point of any instrument with this lens arrangement). Now it can be easily shown that

$$Q = F^2 u - F, \qquad \ldots (81)$$

in which $F$ is the power of the field lens in diopters and $u$ is the distance of the test target from the lens in meters. This relationship shows that the dioptric value, $Q$, of the image distance from the eye is directly proportional to the target distance, $u$. This means that we could place a scale below the axis of the lens to indicate the position of the target, and we could then calibrate this scale linearly in diopters of $Q$. If the lens has a power of 5 diopters and therefore is placed 20 cm from the eye, then each centimeter of change in distance of the target along the axis of the lens changes the vergence of the light seen by the eye by 0.25 diopter. The error in measuring $q$ (and $Q$) from the entrance pupil, rather than from the first principal plane, is negligible.

The angular size of the image corresponding to a target size, $s$, would be given by

$$\tan \alpha = sF. \qquad \ldots (82)$$

That is, the angle subtended by the test target is equal to the product of the target size (meters) and the power of the field lens. It can also be shown that the luminance of the target seen through the lens remains essentially constant for all positions of the target.

The optical principle given here is useful in a number of optical devices such as the lensometer and the oculometer.

## 4. THE LENSOMETER (VERTOMETER AND OTHER DEVICES)

The lensometer is used to measure directly the vertex powers of ophthalmic lenses. The basic arrangement of the several parts is shown in Figure 166, where the power of a converging (plus) ophthalmic lens is shown being measured. Essentially, the *modus operandi* consists in our determining that position of a test target or reticle behind the lens for which the emergent light rays will be parallel.

An illuminated target reticle can be moved back and forth along the axis of the instrument, nearer to and farther from a fixed-field lens. The posterior focal point of this field lens is arranged to coincide with the position of the vertex of the spectacle lens being measured. This spectacle lens is positioned against a fixed lens support with a hole through the center so that the vertex of the lens coincides essentially with the focal point of the field lens. To

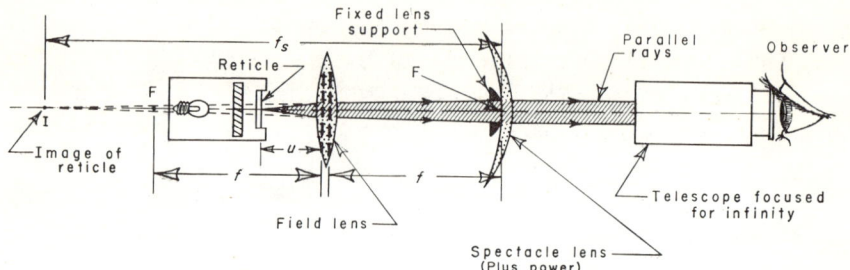

FIGURE 166. Principle of lensometer.

the right of this spectacle lens in the diagram is a fixed-focus telescope focused for an object at infinity—that is, for parallel incident rays. The observer, looking through the telescope, adjusts the position of the target along the axis of the instrument until the image of the target reticle is focused sharply on cross-hairs of the telescope. The rays then entering the telescope must be parallel. Since it is the vertex power of the lens that is to be measured, this measurement is relative to the focal point of the field lens. Hence the Badal principle is used here, for $Q$ (equation 81) is the power corresponding to the distance, $u$, of the reticle target from the field lens. In the figure the spectacle lens has plus power, so its focal point ($I$) is a distance $f_s$ from the lens. The distance $u$ is such, then, that the image of the reticle formed by the field lens coincides with $I$. On the basis of the principle used here, the power, $Q$, is directly proportional to the distance $u$ (equation 81). Thus the dioptric scale indicating the position of the reticle target is a linear scale; that is, equal divisions indicate equal steps of dioptric power. The separation of the field lens and spectacle lens can not be greater than the reciprocal of the power of the strongest plus spectacle lens to be measured.

The angular magnification and the luminance of the image of the reticle target as seen by the observer also remain constant, irrespective of the distance of the target.

Various types of reticles and of reticle housings have been devised also to facilitate measurement of the power and axis of astigmatic lenses, as well as of the prismatic deviation of the lens when centered in the instrument.

## 5. THE STEREOSCOPE

The stereoscope is an optical instrument with which separate pictures can be presented to each of the two eyes at the same time. This instrument is important in ophthalmology and physiologic optics for the study of binocular function.[5] There are many different designs of this instrument, each adapted for special purposes, although all are based upon a very simple principle.

The prototype of all stereoscopes is the mirror stereoscope of Wheatstone (Fig. 167). With this instrument Wheatstone first showed the fundamental difference between the stereoscopic visual perception of spatial depth and the conception of depth and distance arising from secondary visual clues. The diagram is essentially self-explanatory and shows how the targets $A$—$B$ seen by each eye appear by reflection to be superposed in space at $A'$—$B'$. If the spatial image is to be a great distance from the observer, plus lenses must be interposed before the eyes or the mirrors. When used for a near observation distance, each of the targets must be tilted slightly if the two reflected images are to correspond to a single target in space.

A simple stereoscope without mirrors is suggested in Figure 168.

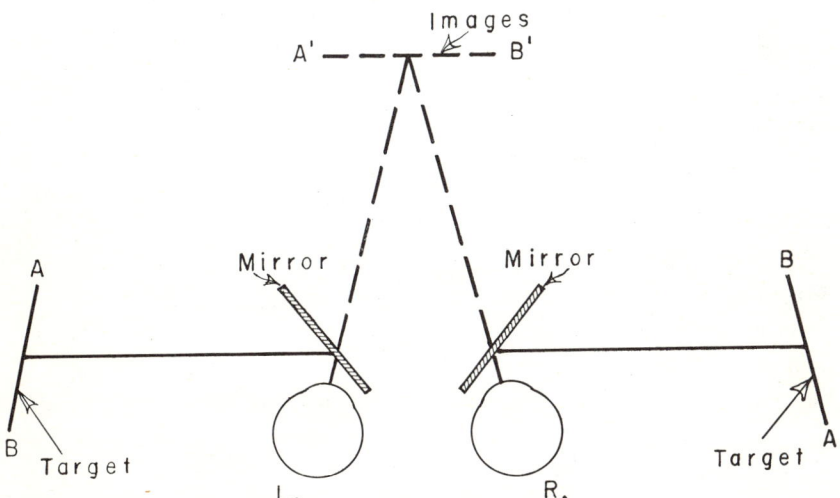

FIGURE 167. Wheatstone reflecting stereoscope.

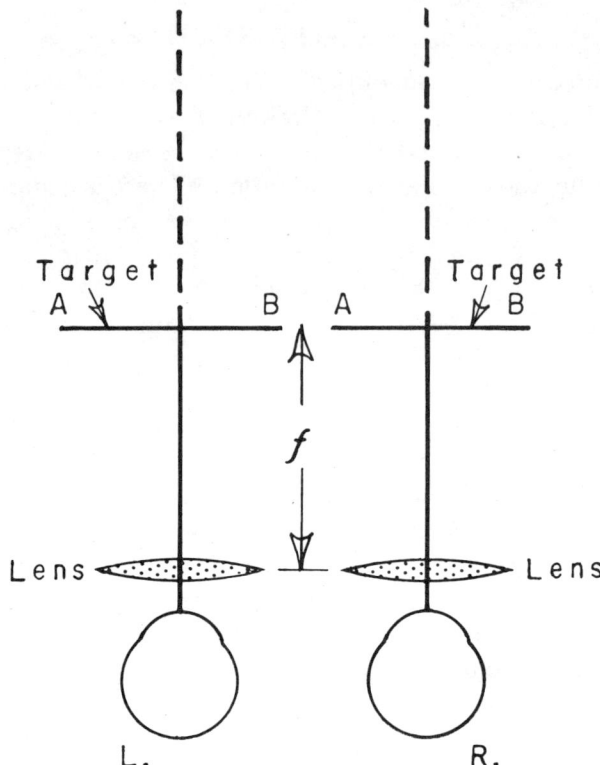

FIGURE 168. Principle of lens stereoscope.

The visual axes of the eyes are more or less parallel, maintained by the use of plus lenses before the eyes to form images of the two targets at infinity. This design provides an instrument of limited use because the nasal portions of the individual targets cannot be wider than half the interocular separation of the eyes. To eliminate the necessity for such small targets, the lenses used can be decentered (Fig. 169) so that each provides effectively a prismatic deviation, base-out. With this design the target sizes can be greatly increased. This instrument is frequently referred to as the *Brewster stereoscope*. Essentially, the design is that of all hand stereoscopes. The stimulus to accommodation from the targets can be increased by moving the targets nearer the eyes. The optical prism arrangement of this stereoscope's design introduces much distortion of the images, as well as chromatic fringes. Care must

# Principles of Certain Ophthalmic Devices 231

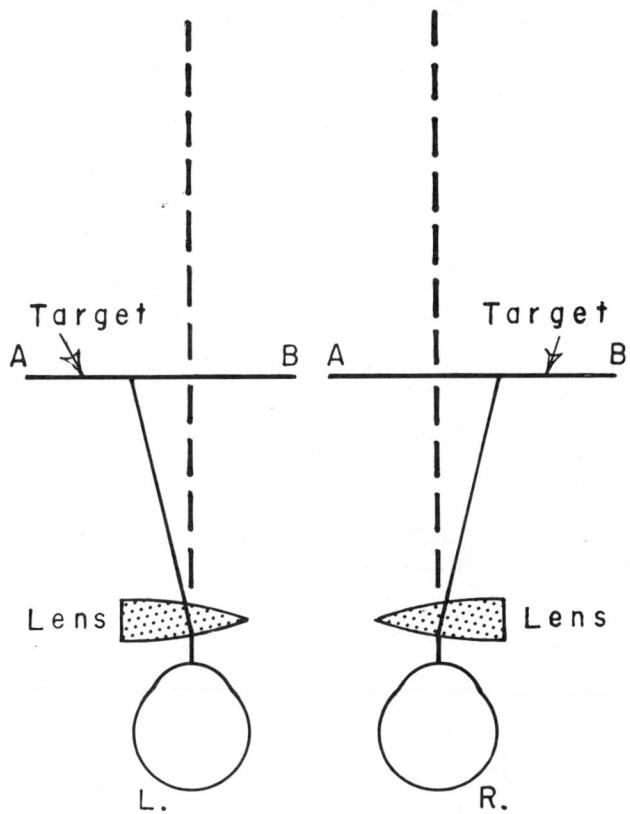

FIGURE 169. Principle of Brewster stereoscope.

be exercised in estimating the degree of accommodation and of convergence of the eyes from a particular position of the targets.

Designed for precision stereoscopy is the *haploscope* of Hering (Fig. 170). Although it is based upon the Wheatstone mirror stereoscope, the haploscope has its mirrors and targets arranged on two arms, each of which rotates about an axis that passes through the center of rotation of one of the eyes. The convergence of the eyes, either symmetric or asymmetric, can be accurately controlled by the converging or diverging of the arms of the haploscope. The stimulus for accommodation can be controlled by adjusting the distance of the targets from the eyes or by inserting suitable lenses in appropriate lens holders in front of the mirrors. The mirrors can also be made 50-50 reflection and transmission

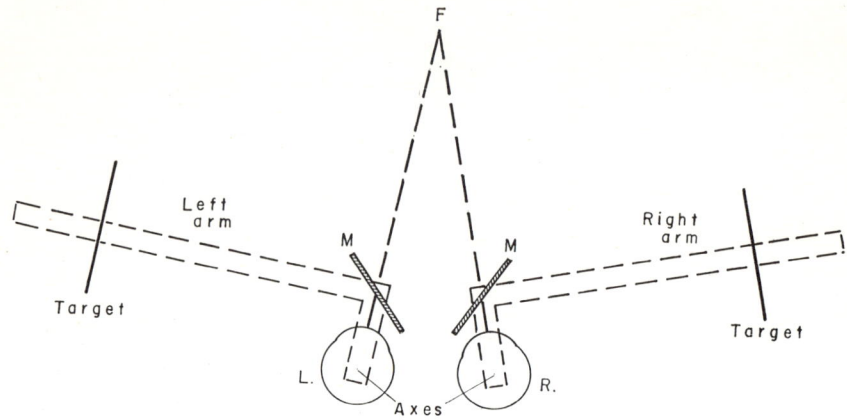

FIGURE 170. Principle of haploscope.

so that actual targets in front of the instrument can be used in conjunction with the images of the separate targets.

The major amblyoscope is an adaptation of the haploscope.

Separate images can also be viewed separately by the two eyes simultaneously by use of *analglyphs* (with red-green filters) and *Polaroid* vectographs.[4] *Free stereoscopy* consists in voluntarily crossing the eyes to view separate targets such as x-ray films: The target to be seen by the right eye is placed to the left of the target to be seen by the left eye.

## 6. THE OCULOMETER[2] (STIGMATOSCOPE[10])

The oculometer is a term sometimes given to an instrument with which the points conjugate to the retinas and the manifest refractive errors of the eyes can be determined by the method of stigmatoscopy (in which the subject adjusts the position of a target so that it appears sharpest) and the Badal principle.

Figure 171 illustrates schematically the essential features of the oculometer which are based on the haploscope. Binocular vision is maintained, and the refractive state of each eye is determined separately. Through 50-50 transmission-reflection mirrors, $M$, the eyes observe a reference chart which provides the stimulus for accommodation, for convergence, and for fusion, at a specified distance. Seen also by reflection from the two mirrors, as if super-

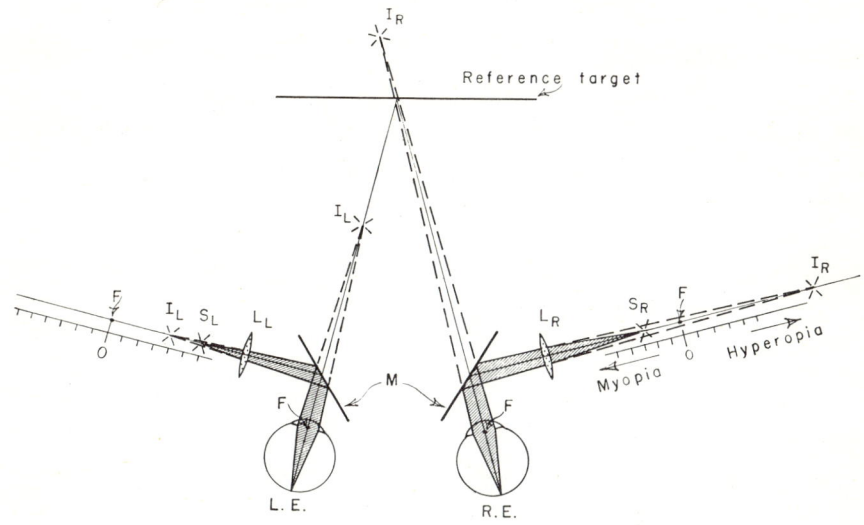

Figure 171. Principle of oculometer.

posed on the target, are the images of point-light sources (or test details of other types) mounted on the right and left arms of the instrument. These point-light sources are seen through field lenses, $L_L$ and $L_R$, whose focal points coincide with the entrance pupils of the right and left eyes, respectively. Thus the angular sizes and luminous intensities of the sources (or targets) remain constant as their positions along the axes of the arms are changed. The actual vergence of the light from the images, and hence the optimal distances of these images, depends upon the position of the sources in relation to the second focal points of the lenses. When the sources are at these focal points, the vergence of the light incident on the eye is zero; that is, the images are as if an infinite distance away. As the sources are moved farther inside the focal length, the vergence of the light incident on the eyes corresponds to that for an object increasingly nearer the eyes. When the sources are outside the focal length, the vergence of the light converges to points behind the eye. Therefore, if an eye is relatively myopic to the distance of the reference chart, the observer adjusting for sharpest and clearest image sets the position of the light source inside the focal length—for example, the left eye in the figure. If the eye is relatively hyperopic to the distance of the reference chart, the observer sets the source closer to the focal point or even beyond that point. According to the Badal principle, the linear scales on the arms can be directly calibrated to indicate the dioptric value, $Q$, of the vergence of the light incident on the cornea. Equal steps of change in vergence correspond to equal distances on the scale.

When the subject has adjusted the light source so that its image on the reference chart appears clearest and sharpest, the corresponding value of $Q$ read from the scale indicates the refractive error in relation to the distance of the chart. Care must be taken in evaluating $Q$ for a very distant chart because most observers tend to be relatively myopic by about 0.25 diopter as tested on the oculometer.[6]

If an eye is astigmatic, the image of the point source of light is likewise astigmatic, so there is a position at which the source appears as a sharp line at one meridian, and another at which it appears as a sharp line at a meridian perpendicular to the first. The corresponding dioptric difference between these two positions, which can be read from the scale, is the measure of the astigmatism.

## 7. THE SCHEINER TEST METHOD

In the previous section, the refractive state of the eye was determined by use of a point source of light whose optical distance was adjusted for sharpest and clearest image. Another method for achieving the same result is that attributed to Scheiner.[9]

Figure 172 shows the principle schematically. A diaphragm (Scheiner disk) with two small holes is centered accurately before the eye so that the pencil of light passing through each hole also passes through the pupil of the eye. When the eye fixates a point source of light, $S$, the observer sees one image if the eye is accurately accommodated for the distance to $S$, and two images if it is not. Such a device has been used in various ways to determine refractive errors.

One modification for this use is the addition of an optical system to project the Scheiner diaphragm at the pupil of the eye. Instead of circular holes, slits can be used. When these slits are offset from each other, the two images on the retina of an illumi-

*Principles of Certain Ophthalmic Devices*      235

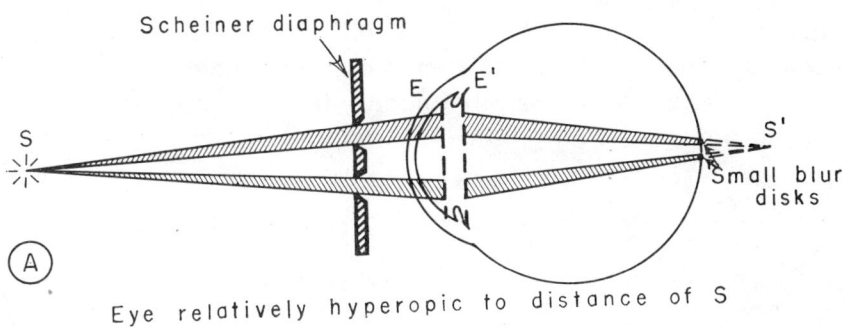

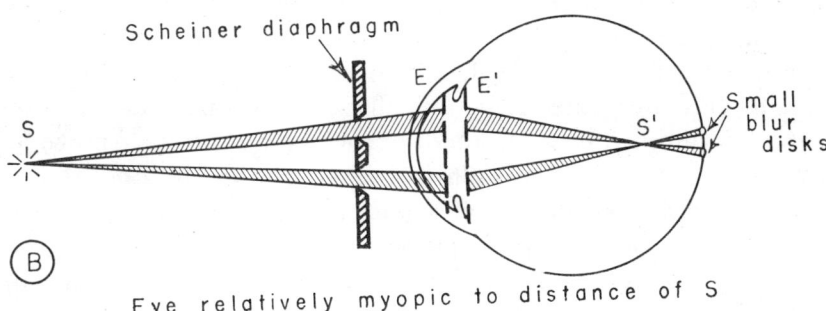

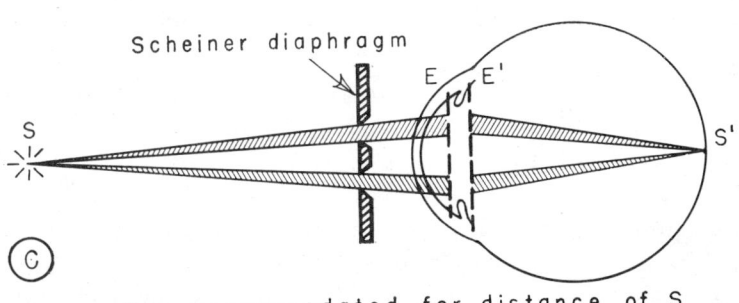

FIGURE 172. Scheiner disk.

nated-line test object parallel to the slits resemble vernier or nonius lines. These line images appear aligned or displaced, depending upon whether the eye is accurately accommodated or whether there is a refractive error. This device has also been adapted for use in the oculometer.

The same principle is used in certain instruments called optometers, which are used for the objective determination of refractive errors. In these, the nonius lines on the retina are observed directly by an ophthalmoscopic technique.

## 8. BIOMICROSCOPY OF THE OCULAR FUNDUS

There are occasions in clinical ophthalmology when it is desirable to observe the fundus of the eye while the corneal microscope of the slit lamp is being used. It is necessary then only to cause an image of the fundus to be formed within the focusing range of the microscope. This can be done by inserting before the subject's eye either a minus or a plus lens of high power.

Figure 173 illustrates schematically the optical arrangements. In the first illustration (A), a lens of high minus power, known as the *Hruby lens*,[8] is used. This planoconcave lens has an inner radius of curvature approximately equal to that of the first surface of the cornea so that it can be placed in contact with the cornea. A virtual image of the fundus is thus produced, which will be in the focal range of the corneal microscope. In the second illustration (B), the same or a similar type of lens is preset in front of the cornea, to produce again a virtual image of the fundus, approximately at the focal point of the lens. This image can also be brought into focus in the corneal microscope. The actual field of the fundus visible in this case is slightly smaller than when the lens is in contact with the cornea.

The third arrangement (C) illustrates the use of a lens of high plus power similarly set in front of the cornea—this time to form a real, inverted image of the fundus, which also can be brought into focus in the corneal microscope. This arrangement is sometimes helpful in cases of high myopia.

## REFERENCES

1. BADAL, M.: Optomètre métrique international: pour la mesure simultanée de la réfraction et de l'acuité visuelle méme chez les illettrés. *Ann Oculist (Paris)*, 75:101-117, 1876.
2. BRICKLEY, P. M., and OGLE, K. N.: Residual accommodation under homatropine-cocaine cycloplegia. *Amer J Ophthal*, 36:649-660, 1953.

# Principles of Certain Ophthalmic Devices

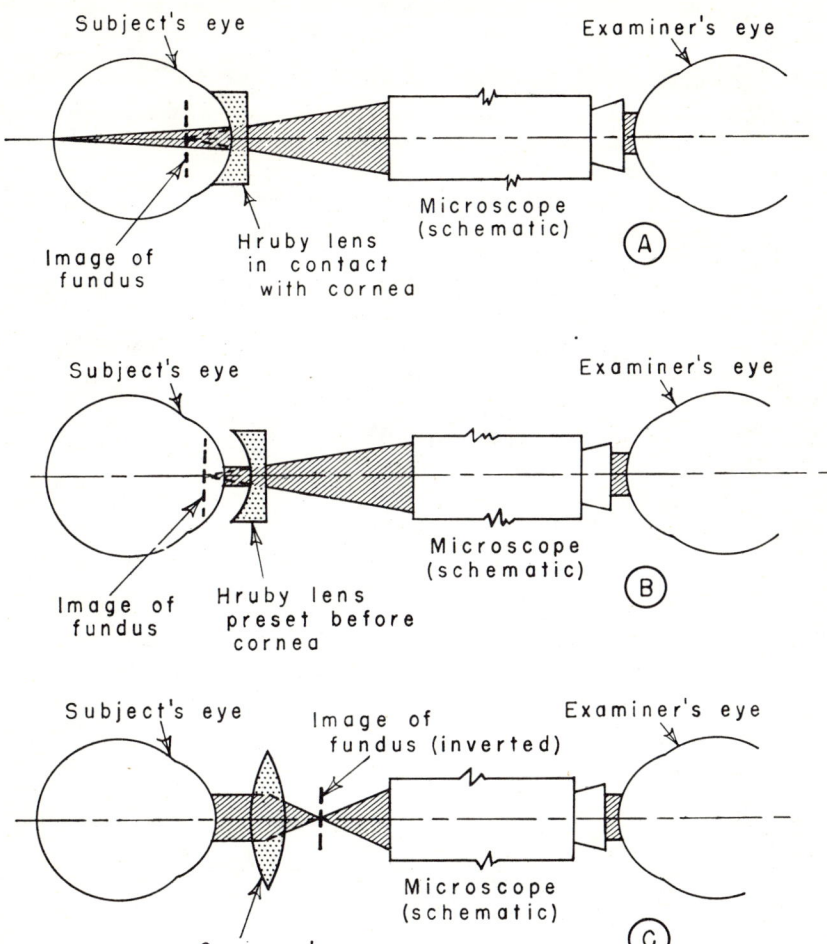

FIGURE 173. Optical principles used in biomicroscopy of the ocular fundus: (A) Hruby lens in contact with cornea; (B) Hruby lens preset; and (C) convex lens preset.

3. von Helmholtz, H.: *Helmholtz's Treatise on Physiological Optics*, trans. from the 3rd German ed., J. P. C. Southall (Ed.). Lancaster, The Optical Society of America, 1924, vol. 1, pp. 249-250. Reprinted 1964, New York, Dover.
4. Land, E. H.: Vectographs: images in terms of vectorial inequality and their application in three-dimensional representation. *J Opt Soc Amer*, 30:230-238, 1940.

5. Linksz, A.: The stereoscope as an orthoptic instrument. *Arch Ophthal (Chicago)*, *26:*389-407, 1941.
6. Ogle, K. N., Imus, H. A., Madigan, L. F., Bannon, R. E., and Wilson, E. C.: Repeatability of ophthalmoeikonometer measurements. *Arch Ophthal (Chicago)*, *24:*1179-1189, 1940.
7. Pascal, J. I.: *Studies in Visual Optics.* St. Louis, Mosby, 1952, pp. 155-224.
8. Rotter, Hans: Technique of biomicroscopy of the posterior eye. *Amer J Ophthal*, *42:*409-415, 1956.
9. Scheiner, C., quoted by Southall, J. P. C.: *Introduction to Physiological Optics.* New York, Oxford U. P., 1937, p. 84.
10. Wald, G., and Griffin, D. R.: Change in refractive power of human eye in dim and bright light. *J Opt Soc Amer*, *37:*321-336, 1947.

Chapter XIV

## ILLUMINATION

From time to time the ophthalmologist will be concerned with problems of illumination, and it is important that he be familiar with at least a few of the basic concepts. At times, these concepts seem difficult to grasp, and they are easily forgotten unless they are used often. For the most part, the problems have to do with the need of expressing light intensities in quantitative terms—that is, in photometric units.[1] These units, which depend upon the response of the eye to radiant energy, are necessarily based upon arbitrary standards. Also, they are frequently derived from over-simplifications of certain theoretic notions and often are only approximations. The use of both English and metric systems in the definition of units has led to additional confusion.

It is convenient and usually sufficiently accurate to treat a source of radiant energy as though it were a point source. As such, this source is said to emit radiant energy equally in all directions. That part of the radiant energy which is effective in producing the visual sensation of light is called luminous energy. The brightness and color of the light depend not only on the intensity of the source but also on the spectral distribution of that radiant energy (see pages 19ff). The dependence of the brightness of the visual sensation on wavelength is due to the fact that the eye does not respond equally to equal intensities of radiant energy at all wavelengths. Since the eye is an integrating device, the brightness of the light may be thought of as the sum of the responses at each wavelength.

There are essentially four aspects of illumination to keep in mind, each of which is based upon a rather simple notion.[5] These four aspects are concerned with how to specify the following:

1. The intensity of the source of luminous energy.

2. The intensity of the light (luminous flux) falling on a given surface.
3. The intensity of the light or luminous energy being reflected from or transmitted by a given surface.
4. The intensity of the light falling on the retina, particularly as determined by size of the pupil.

## 1. SOURCE INTENSITY

In order to quantify the luminous intensity of a source, some arbitrary but standard unit must be agreed upon. This unit is called the candle, a unit now antiquated because the standard—depending on the sperm oil candle—is difficult to reproduce. Although other equivalent standards are now used, the unit is still frequently called the candle. The standard unit today (the candela) is defined as one sixtieth of the luminous intensity of a square centimeter of a black-body radiator at the freezing point of platinum (2,047 degrees Kelvin).

A measure of the light-producing power of the source is the luminous output or luminous flux. Thus, although the number of candles may adequately indicate the intensity of the luminous source itself, a unit is needed to express the intensity of the *luminous flux* being emitted in any one direction or, more accurately, within a given solid angle.* In Figure 174, the same amount of luminous flux is passing per second through each of the areas $A_1 \ldots A_n$ because each subtends the same solid angle. The unit to describe this luminous flux is the *lumen*. The lumen is described as the luminous flux proceeding from a light source in a unit solid angle—that is, a solid angle specified by a portion of a sphere with its center at the source—having an area equal to the square of the distance from the source. Thus, in Figure 175, where the light source is indeed one candlepower, a spherical surface area of one square foot at a distance of one foot, or an area of one square meter at a distance of one meter, defines the unit solid angle enclosing a luminous flux of one lumen. The total number of

---

*A solid angle, ω, is defined as the cone-shaped section of a solid sphere with apex at the center, subtending an area, $A$, on the surface of the sphere. The solid angle, ω, is equal to $A/r^2$, in which $r$ is the radius of the sphere.

# Illumination

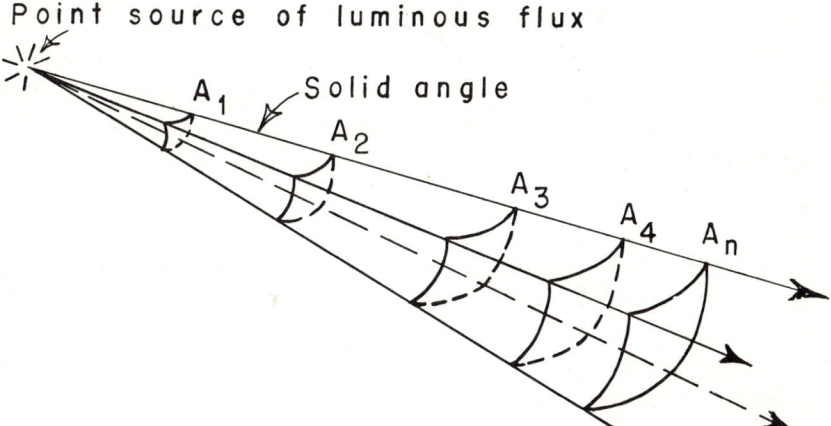

FIGURE 174. Illustration of luminous flux from a point-light source within a given solid angle.

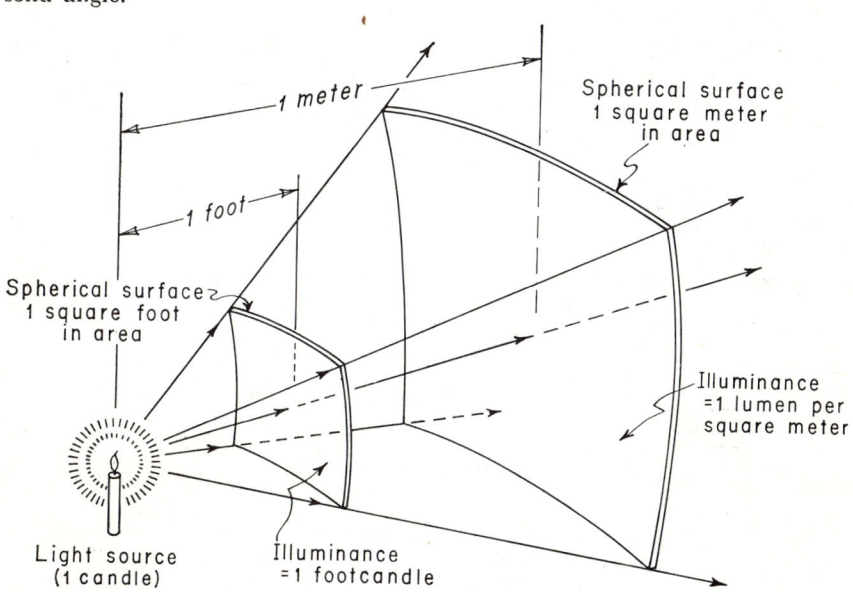

FIGURE 175. Lumen of luminous flux illustrated by the unit solid angle from a light source.[3] (Modified by permission of the General Electric Company.)

lumens of luminous flux emitted by an ideal point source of light of one candle is $4\pi$ (or 12.56) lumens, since the total area of a sphere is $4\pi r^2$.

The luminous flux (measured in lumens) depends on the magnitude of the radiant flux (watts-seconds) being emitted by the source. The *ratio* of the luminous flux of a light source in lumens to the radiant energy in watts is called the *luminous efficiency* of the source (lumens/watt). An example of incandescent electric lamps is a one-hundred-watt bulb which emits about 1,700 lumens: hence the luminous efficiency is between fifteen and twenty lumens per watt. An example of fluorescent lamps is a forty-watt tube which emits approximately 2,100 lumens; hence the approximate luminous efficiency is about fifty lumens per watt, which includes allowance for the power loss in the transformer or ballast. This efficiency varies with the wattage of the lamp.

## 2. ILLUMINANCE

Returning to Figure 175, it is clear that the *density* (lumens per unit area) of the luminous flux on the surface depends upon the distance of the surface. Thus the luminous-flux density of the light falling on the surface at one foot is much greater than that at one meter.

A surface upon which light—or more precisely, luminous flux—falls is said to be *illuminated*. The magnitude (intensity) of that illumination—called the *illuminance*—is the luminous flux (lumens) falling on the surface per unit area of that surface. In Figure 175, the illuminance on the first surface is one lumen per square foot or, as this unit is defined, one footcandle. On the second surface the illuminance is one lumen per square meter (sometimes called the *lux*). Since there are 929 square centimeters in a square foot, an illuminance of one lumen per square meter corresponds to 0.0929 footcandles, and an illuminance of one footcandle corresponds to 10.76 lumens per square meter. Thus a given radiant flux provides ten times more illuminance at one foot than at one meter.

The illuminance, or luminous flux falling on a surface, can be measured by an illuminometer. The hand-sized photovoltaic-cell illuminometers (Weston, for example) used in this country are calibrated to read in footcandles. A General Electric light meter is illustrated in **Figure 176**.

*Illumination*

FIGURE 176. Color- and cosine-corrected G.E. Type-213 light meter featuring a logarithmic scale.[3] (By permission of the General Electric Company.)

The inverse-square law of illumination is almost self-evident from the preceding discussion: The illuminance on a surface is inversely proportional to the square of its distance from the source. Certainly if $L_o$ is the illuminance on a surface (in footcandles) at a distance $r_o$ from a luminous source, then the illuminance $L$ on another surface (also in footcandles) at a distance $r$ from the source is given by

$$\frac{L}{L_o} = \frac{r_o^2}{r^2}. \qquad \ldots (83)$$

If the second surface were slanted at an angle $\theta$, then the illuminance $L$ would be further reduced by the factor $\cos \theta$.

### 3. LUMINANCE

The light reflected from an illuminated surface, or emitted from an extended luminous source (and it is only this light flux to which the eye usually responds), is called the *luminance*. The subjective sensory equivalent to luminance is *brightness*.

For a uniformly illuminated surface which reflects light equally in all directions, the intensity of luminous flux reflected (or emitted) per unit area of that surface is the luminance. Here again, the unit used depends upon the unit of area selected as reference: square meter, square foot, or square centimeter. "A theoretical perfectly diffusing surface emitting or reflecting [luminous] flux at the rate of one lumen per square foot would have a [luminance] of one *foot-lambert* (*fL*) in all directions."[4] If the rate of luminous flux is one lumen per square centimeter, the luminance is one *lambert* (*L*). Conversely, each square centimeter, having a luminance of one lambert, emits a luminous flux of one lumen. Now the actual magnitude of luminance corresponding to this unit is far too high for general use, so it is convenient to use the *millilambert* (*mL*), which is 1/1,000 of the lambert. For very low luminances, corresponding to scotopic visual conditions, even the *microlambert* ($\mu L$), which is 1/1,000,000 of a lambert, can be used.

If the unit of area is one square meter, the luminance unit is one lumen per square meter or one meter-lambert. If the unit of area is one square foot, the unit of luminance is one lumen per square foot, or one *foot-lambert*. Since there are 929 square centimeters in a square foot, one foot-lambert is equal to 1.076 millilamberts, so these two units are approximately equal. Occasionally the luminance is expressed in candles per square meter, which is equal to meter-lambert divided by $\pi$ (a relationship based on theoretical grounds).

Luminance can be measured by several instruments such as the Macbeth illuminometer or certain photoelectric devices or photovoltaic cells such as the exposure meter.

To give some definite ideas about common magnitudes of luminance, Figure 177 provides an approximate scale, based on everyday experience.

## 4. RETINAL ILLUMINANCE

The brightness of a surface or object is related in a general way to the light stimulus—that is, to the illuminance of the image

*Illumination* 245

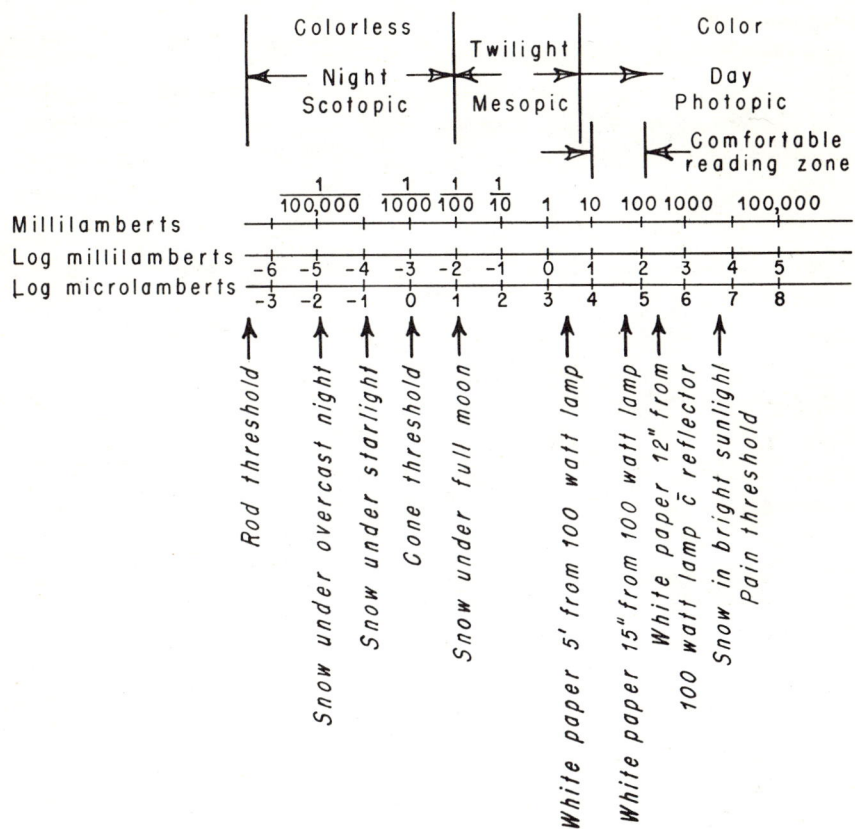

FIGURE 177. Luminance scale related to everyday experience.

falling on the retina. Even if the luminance of an object being observed remains constant, the visual stimulus (retinal illuminance) varies with the area of the pupil of the eye: In particular, it increases proportionally with the square of the diameter of the entrance pupil.

For comparison of the intensity of visual stimuli with different sizes of pupil, a unit for retinal illuminance is needed. The unit most frequently used in this country is called the *troland*. The troland is defined as "the retinal illuminance produced by an image of an object the luminance of which is one candle per square meter for an area of the entrance pupil of one square millimeter." Since the candle per square meter is equal to $10/\pi$

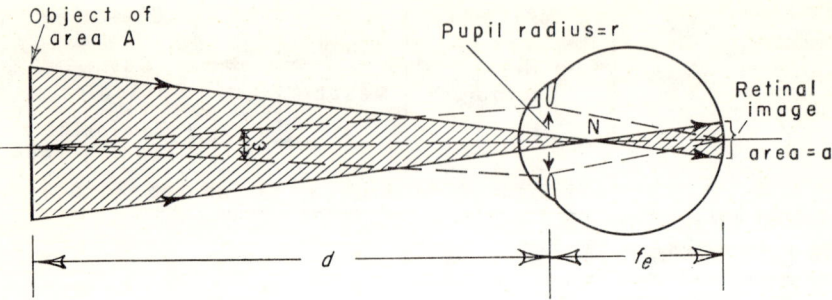

Figure 178. Illustration of derivation of illuminance of the retinal image.

or 3.18 millilamberts, and a pupil of radius $r$ has an area of $\pi r^2$, this definition leads to $E$ (trolands) $= 10\ Br^2$, in which $B$ is the luminance of the object surface in millilamberts.[2] Thus, for a 3-mm pupil, $E = 22.5\ B$, or $\log E = \log B + 1.35$. These relationships provide us with simple rules for converting millilamberts to trolands or vice versa, on the assumption of a pupil 3 mm in diameter.

Because of the Stiles-Crawford phenomenon (which is that light passing through a given area in the peripheral parts of the pupil produces a sensation of less brightness than does light of the same intensity passing through an equal area in the center of the pupil), the use of any unit such as the troland is beset with difficulties. Although corrections can be applied, this effect is by no means the same in all eyes.

The illuminance of the retinal image of a given object which has constant luminance is independent of the distance of the object, for it depends only upon the luminance of the object itself and upon the size of the pupil. This fact will be evident from the following: In Figure 178, let the eye be observing an object having an area, $A$, at a distance, $d$, from the eye. This object then has a certain luminance (lumens per square centimeter, or lamberts), $B$. If the radius of the pupil is $r$, then the solid angle, $\omega$, subtended by the pupil (its area $= \pi r^2$) at a point on the object is $\omega = \pi r^2/d^2$. Then the luminous flux entering the eye from each point on the object is $B\omega = \pi r^2 B/d^2$. The total luminous flux for the entire object is $(B\omega)$ multiplied by the area of the object, $A$. If we can assume that no light is lost in passing through the ocular media, this gives the total luminous flux in the corresponding image on the retina. If $a$ is the area of the image of the object, then from similar triangles one can write the ratio of the areas of image and object as

$$a/A = (f_e/d)^2,$$

in which $f_e$ is the focal length of the eye. The illuminance of the retinal image—lumens per unit area—is obtained by solving the above equation for $a$ and then dividing the total light flux, $AB\omega$, in the image by it. So the illuminance, $L$, of the retinal image is

$$L = \pi r^2 B / f_e^2. \qquad \ldots (84)$$

This equation states that the illuminance of the retinal image, and therefore the visual stimulus for brightness (other factors being constant), depends only upon the luminance of the object, the size of the pupil, and the focal length of the eye. This relationship applies to all optical systems.

A given visual acuity chart with a constant illuminance appears just as bright at a near distance as at a far one. The light flux entering the eye decreases with the square of the distance to the chart, but the area of the retinal image, being inversely proportional to the object distance, decreases with the square of that distance. Thus the luminous flux per unit area of the image, which is what determines the visual stimulus for the perception of brightness, remains constant.

## f-Number

The above relationship for retinal illuminance of an optical image can be written in general terms as $L = [\pi B/4][2r/f]^2$. In photographic optics, the ratio $f/2r$ (the ratio of the focal length to the diameter of the aperture) is called the f-number, and it is also referred to as the speed of the optical system. The illuminance of the image, therefore, is inversely proportional to the square of the f-number: The larger the f-number, the lower the illuminance and hence the lower the speed. The apertures or iris openings of photographic lenses are calibrated according to f-numbers in steps such that the illuminance of the image for one opening is just half that for the preceding opening—that is, in steps of ½ (f-number)². Thus one can compare the speeds of all photographic lenses in terms of the illuminance of the images produced. In the case of the human eye with a pupil diameter of 4 mm, the f-number is about 4.3.

## Lighting Standards

Lighting and illuminating engineering has rapidly become a scientific discipline in its own right. It has certainly become a necessary part of architectural design and interior decorating. There has been a great deal of research into the problems of lighting levels, lighting fixtures, and lighting arrangements required for comfort and efficiency in the performance of visual tasks.

The visibility of a visual task, whether reading, sewing, machine assembly, or inspection, depends upon the size of the details to be discriminated and on the contrast between the details and their backgrounds. Low contrasts and small details require higher illuminances; high contrasts and large details require lower illuminances; both situations depend on the light adaptation of the eyes. In addition to control of the quantity of illumination in given surroundings, control of its quality is needed. By the latter is meant, among many things, the degree to which the light is diffused through control of reflection from walls, ceilings, furniture surfaces, and so on; the distribution of general lighting and the need for supplementary lighting; the absence of glare; and the use of color. As a general rule, the difference in illuminances (considering the reflection factors of the various surfaces to be about the same) of a given visual task and the immediately adjacent surfaces should be about one to one-third, and the difference between the task and more remote surfaces (desk surface and floor) should be about one to one-tenth.

For engineering problems the illuminance is usually measured with a light meter or similar instrument on a working plane about thirty inches (76 cm) above the floor. Since it is the luminances with which we are most concerned, it is important to know approximately the reflectance factors in estimating the quantity of illumination needed on this surface.

Examples of currently recommended illuminance requirements for various interiors and visual tasks are given in the following table (selected from the *IES Lighting Handbook*). These levels are given in footcandles (about ten times one lumen per square meter)

| Area | Footcandles |
|---|---|
| *Offices* | |
|   Routine tasks | 100 |
|   Accounting | 150 |
| *Schools* | |
|   Rooms, general | 30 |
|   Reading printed material | 30 |
|   Reading pencil writing | 70 |
|   Drafting tables | 100 |
| *Corridors and halls* | 20 |
| *Elevators* | 20 |
| *Hospitals* | |
|   Operating table | 2,500 |
|   General | 30 |
| *Libraries* | |
|   Studies | 70 |
|   Ordinary reading | 30 |
|   Stacks | 30 |
| *Residences* | |
|   Range, work surfaces | 50 |
|   Ironing board | 50 |
|   Reading, etc. | 30 |
|   Music score | 30–70 |
|   Sewing | |
|     Light fabrics | 50 |
|     Dark fabrics (low contrast) | 200 |
|   Living room (relaxing, conversing) | 10 |
| *Basketball courts* | 50 |
| *Parking areas* | 1–2 |

For detailed information on such applications of lighting the reader is referred to the latest edition of the *IES Lighting Handbook* and to readily available pamphlets.

## EXERCISES

1. On a photometer bench, a standard lamp of 40-candles intensity and a lamp of unknown candlepower are separated by 300 cm. Between them, the photometer shows that the luminances of the two are equal at a point 100 cm from the standard lamp. What is the intensity or candlepower of the unknown lamp?
2. The illuminance on a desk top from an approximately point-light source directly above the desk at a distance of 2 feet is 50 lumens per square foot. What is the intensity of the light source?
3. If the reflectance of a white paper on the desk in problem 2 is 0.80, what is the luminance of the paper in millilamberts?

4. The luminance of a visual-acuity chart is 80 millilamberts. What is the retinal illuminance (trolands) of the image, if the pupil is 5 mm in diameter?
5. If the *f*-number of the eye is 4.3 when the pupil diameter is 4 mm what is the f-number when the pupil is 8 mm? How does the f-number compare with those for camera lenses?

## REFERENCES

1. COMMITTEE ON COLORIMETRY, OPTICAL SOCIETY OF AMERICA: *The Science of Color.* New York, Crowell, 1953, p. 229.
2. DE GROOT, S. G., and GEBHARD, J. W.: Pupil size as determined by adapting luminance. *J Opt Soc Amer, 42:*492-495, 1952.
3. General Electric Company: *Light Measurement and Control,* Bulletin TP-118. Cleveland, General Electric Company, pp. 5-6.
4. Illuminating Engineering Society: *IES Lighting Handbook,* 4th ed. New York, Illuminating Engineering Society, 1966.
5. SEARS, F. W.: *Principles of Physics, III: Optics.* Reading, Addison-Wesley, 1945, p. pp. 276-295.

*Appendix A*

# RECOMMENDED READING

THE following books are recommended for collateral and supplementary reading and study.

## GENERAL

Chapters on optics in various physics textbooks, such as the following:

BENNETT, A. G., and FRANCIS, J. L.: Visual optics. In Davson, H: *The Eye*. New York, Academic, 1962, vol. 4, pp. 1-210.
DITCHBURN, R. W.: *Light*, 2nd ed. New York, Interscience, 1963.
HARDY, A. C., and PERRIN, F. H.: *The Principles of Optics*. New York, McGraw, 1932.
JENKINS, F. A., and WHITE, H. E.: *Fundamentals of Optics*, 2nd ed. New York, McGraw, 1950.
SEARS, F. W.: *Principles of Physics, III: Optics*. Reading, Addison-Wesley, 1945.
WHITE, HARVEY E.: *Modern College Physics*, 3rd ed. Princeton, Van Nostrand, 1956, pp. 374-474.

## OPHTHALMIC AND VISUAL OPTICS

COWAN, ALFRED: *Refraction of the Eye*, 3rd ed. Philadelphia, Lea & F., 1948.
DUKE-ELDER, W. S.: *Textbook of Ophthalmology*. St. Louis, Mosby, 1944, vol. 1, pp. 659-818.
EMSLEY, H. H.: *Visual Optics*, 4th ed. London, Hatton Press, 1946.
LINKSZ, ARTHUR: *Optics*. Vol. 1 in Physiology of the Eye. New York, Grune, 1950.
PASCAL, J. I.: *Selected Studies in Visual Optics*. St. Louis, Mosby, 1952.
SHEARD, CHARLES: Ophthalmic optics with applications to physiological optics. In Glasser, Otto: *Medical Physics*. Chicago, Year Bk., 1950, vol. 2, pp. 617-658.

SHEARD, CHARLES: Ophthalmic optics with applications to physiological optics. In *The Sheard Volume: Selected Writings in Visual and Ophthalmic Optics*, 1st ed. Philadelphia, Chilton, 1957, pp. 291-400.

SOUTHALL, J. P. C.: *Mirrors, Prisms and Lenses: A Text-book of Geometrical Optics*, 3rd ed. New York, Macmillan, 1933. Reprinted 1964, New York, Dover.

## OLDER OR OUT-OF-PRINT TEXTS

BOEDER, PAUL: *An Introduction to the Mathematics of Ophthalmic Optics*. Fall River, The Distinguished Service Foundation of Optometry, 1937.

EMSLEY, H. H., and SWAINE, WILLIAM: *Ophthalmic Lenses*, 2nd ed. London, Hatton Press, 1932.

LAURANCE, LIONEL: *General and Practical Optics*, 3rd ed. London, School of Optics, 1920.

*Appendix B*

# ANSWERS TO THE EXERCISES

## CHAPTER II

1. 3 feet
6. $v = 3.85$ mm; $I = 1.54$ mm
7. $O = 37.5$ cm
8. $R = 7.6$ mm
9. $v = -3.33$ cm; $I = -0.333$ cm
10. $R = -80$ cm; $f = -40$ cm; $M = 2.0$
11. $R = -40$ cm
12. 56 mm in front of the eye; No
13. (a) The image would be at $v = -33.3$ cm, and the height $h' = 3.33$ cm
    (b) The change in distance would be 1.3 cm, and the change in height would be 0.51 cm

## CHAPTER IV

1. 8.24 prism diopters
2. Prismatic deviation = 5 prism diopters; prismatic deviation 5 prism diopters
3. Prismatic deviation = 5.4 prism diopters; $\phi = 158.2°$
4. $H = 4.3$ prism diopters; $V = 2.5$ prism diopters

## CHAPTER V

1. 1.00, 3.00, 2.50, 4.00, 2.00, and 10.00 diopters, respectively
2. $F = 2.50$ diopters
3. $F = 8.00$ diopters
4. $v = 20$ cm; $F = 7.50$ diopters (also, $-2.50$ diopters)
5. $v = 20$ cm; $I = 3$ cm
6. $v = -9.09$ cm; $I = 1.36$ cm; virtual
7. $v = -1,000$ mm; $I = 24$ cm; virtual

## CHAPTER VI

1. $A = 1.0375$; $3.75\%$
2. $A = 6.00$
3. $20/120$
4. $F = 25.00$ diopters; $h = 5.5$ cm
5. $+185$ mm; $-140$ mm; $+85$ mm; $-81.6$ mm
6. $V_o = +4.00$ diopters (rigorous method); $V_o = +3.75$ diopters (formula)
7. Effective power $= -10.80$ diopters (formula)
8. $V_o' = +14.38$ diopters (formula); $V_o' = +14.71$ diopters (rigorous method)
9. $26.8$ cm

## CHAPTER VII

1. (a) $F = 12.50$ diopters
   (b) $f = 8.0$ cm
   (c) $a = +8$ mm; $b = -5$ mm
   (d) $f_b = 7.5$ cm; $f_a = 7.2$ cm
   (e) $9.4$ cm
   (f) $M = -0.242$
2. (a) $F = 1.25$ diopters
   (b) $f = 80$ cm
   (c) $a = -4.0$ cm; $b = -5$ cm
   (d) $f_b = 75$ cm; $f_a = 84$ cm
   (e) $v + b = -70.5$ cm
   (f) $M = 1.82$
3. $s = 2$ cm; $A = 1.25$; No

## CHAPTER VIII

1. $D = 1.25, 2.00, 5.00, 7.14$, and $25.00$ diopters, respectively
2. $D = 8.35$ diopters
3. $F = 1.67$ diopters
4. $43.2$ diopters
5. $2.95$ cm; $I = -0.79$ cm
6. $F = 12.3$ diopters; $a = 1.6$ mm; $b = -6.5$ mm
7. $18$ cm

*Appendix B*

## CHAPTER IX

1. $V_o = -5.39$; $F = -5.29$ diopters; $V_o = -5.13$

2. 

|     | True power ($F$) (diopters) | Vertex power ($V_o$) (diopters) | Vertex power ($V_o'$) (diopters) (lenses reversed) |
|-----|---|---|---|
| (a) | −1.95 | −1.96 | −1.94 |
| (b) | +6.11 | +6.27 | +6.04 |
| (c) | +7.40 | +7.74 | +7.21 |
| (d) | −1.89 | −1.92 | −1.87 |

3. (a) 0.05 diopter
   (b) 0.13 diopter
   (c) 0.27 diopter
   (d) 0.75 diopter
4. (a) −8.51 diopters
   (b) −3.64 diopters
   (c) −2.48 diopters
5. $A = 1.0309$; $D_2 = -10.30$ diopters; $m\% = 3.0\%$
6. 0.39 mm/diopter
7. (a) 1.8 mm
   (b) 21 cm
   (c) 63 cm
8. $I = 0.855$ mm
9. $I = 0.005$ mm $= 5$ microns $(5\mu)$

## CHAPTER X

1. $D = +3.76$ diopters and $D_\perp = +6.00$ diopters; astigmatism $= 2.24$ diopters
2. +2.00-diopter sphere ⌒ +2.00-diopter cylinder
   (or)
   +4.00-diopter sphere ⌒ −2.00-diopter cylinder
3. 33.3 cm and 16.6 cm; separation $= 16.6$ cm; spherical equivalent $= 4.50$ diopters; 22.2 cm
4. 16.6 cm and 11.1 cm; separation $= 5.5$ cm
5. (a) 10.72 diopters and 6.89 diopters
   (b) 6.89 diopters and 2.98 diopters
6. (a) +3.50 diopters
   (b) −6.50 diopters
   (c) 0.00 diopters
   (d) −1.00 diopter

7. (a) $+1.50 \times 90°$ and $+1.50 \times 180°$
   (b) $+3.00 \times 90°$ and $+1.00 \times 180°$
   (c) $+1.87 \times 90°$ and $+0.62 \times 180°$
   (d) $+0.75 \times 90°$ and $+2.25 \times 180°$
   (e) $-0.75 \times 90°$ and $-2.25 \times 180°$
8. (a) 0.75 prism diopter base-in; 1.20 prism diopters base-down
   (b) 1.51 prism diopters base-in; 0.80 prism diopter base-down
   (c) 0.94 prism diopter base-in; 0.50 prism diopter base-down
   (d) 0.37 prism diopter base-in; 1.80 prism diopters base-down
   (e) 0.37 prism diopter base-out; 1.80 prism diopters base-up
9. (a) $+2.00$ sph. $\circ$ $+2.00$ cyl. $\times$ 123°
   (b) $-8.00$ sph. $\circ$ $+3.00$ cyl. $\times$ 70°
   (c) $-1.00$ sph. $\circ$ $-3.00$ cyl. $\times$ 180°
   (d) $-1.00$ sph. $\circ$ $+3.00$ cyl. $\times$ 90°
   (e) $-2.50$ sph. $\circ$ $+2.50$ cyl. $\times$ 135°
   (f) $+4.00$ sph. $\circ$ $-4.00$ cyl. $\times$ 70°

## CHAPTER XI

1. Blue
2. 1 meter; 3 feet (approximately)
3. Maximum myopia = ½ depth of focus = 0.64 diopter
4. $(d - p)$ = (200 cm − 50 cm) = 150 cm
5. Resolving power = 0.40 diopter; Snellen equivalent = 20/50
6. 21.8 mm

## CHAPTER XII

1. $A$ = 1.175 or 17.5%; $\alpha'$ = 35.25 degrees; prismatic deviation = 5.25 degrees = 9.1 prism diopters, base-left
2. $A$ = 0.87 or $-13\%$; $\alpha'$ = 26.1 degrees; prismatic deviation = 3.9 degrees = 6.8 prism diopters, base-right
3. Prismatic deviation = 15.9 prism diopters, base-in
4. $+8.65$ prism diopters; $-8.65$ prism diopters; vertical divergence = 17.3 prism diopters
5. Vertical divergence = 15.9 prism diopters (rigorous) and 17.3 prism diopters (approximate)
6. $M$ = 1.05 or 5%
7. $\partial v$ = (0.29) (5.0) $(I)$ = 1.45 arc degrees; excyclotorsional

8. Total prismatic deviation = 2.0 prism diopters, base-out (prismatic deviation of each lens = 1.0 prism diopter); each lens should be decentered $d$ = 2 mm in (nasally).

## CHAPTER XIV

1. $I$ = 160 candles
2. $I$ = 200 candles
3. $B$ = 43 millilamberts
4. $L$ = 5,000 trolands
5. $f$-number = 2.15

# INDEX

## A

Abbé refractometer, 54
Aberration, 29, 181
  chromatic, 72, 182
    of human eye, 183
  negative, 182
  spherical, 46, 129, 181, 182
Absorption, 22ff
  coefficient of, 24
Accommodation, 103
Afocal lenses, 148
  Galilean, 148, 197
Afocal magnifying lenses, 206
  meridional, 207
  overall, 207
Afocal system, 122
Allowance factor, 147
Ametropia, 104
  axial, 204
  hyperopic, 107
  myopic, 104
  refractive, 202
Analglyphs, 232
Angstrom units, 16
Aniseikonia, 206
Apex, 64
  angle, 64, 67
Aphakia, 215
Aphakic eye, 202
Aplanatic system, 29
Aspheric curve, 182
Astigmatic imagery, 162
Astigmatic interval, 165
Astigmatism, 162
  from tilted lenses, 187
  irregular, 162
  of oblique incidence, 185, 187
Astronomical telescopic system, 123
Axis, geometric, 168
Axis, optic, 37, 85

## B

Badal principle, 226

Barrel distortion, 189
Base-apex meridian, 65
Base curve, 132
Beam of light, 27
Beer's law, 25
Biomicroscopy of ocular fundus, 236
Biprism, 80
Bitoric lens, 207
Black-body radiation, 19, 240
Blur disk, 190, 193
Blurred imagery, 190
Bouguer's law, 24
Brewster stereoscope, 230
Brightness, 244

## C

Candle, 240
Cardinal points, 154
  of schematic eye, 157
Catoptric images, 46
Caustic curve, 46, 57
Caustic surface, 46, 57, 181
Centrad, 70
Central diffraction disk, 193
Chromatic difference in magnification, 190
Circle of least confusion, 165
Coefficient, molecular absorption (extinction), 25
Coefficient of absorption, 24
Color temperature, 21
Coma, 184
Conjugate points, 29, 91
Conoid, Sturm's, 166
Corning glass filters, 23
Critical angle, 53
Cross-cylinder test lens, 176
Curvature of field, 187
Cylinder at oblique axis, 171
Cylinders, geometric, 168
  obliquely crossed, 173
Cylindric lenses, components of, 171
  scissors effect of, 177

## D

Density, 23
Depth of field, 194
Depth of focus, 193
Diffraction, 6, 13
Diffraction grating, 14
    concave, 14
Diffraction pattern, 13
Dioptric power, 93
    of a surface, 130
Dioptric constants of the schematic eye, 157
Direct ophthalmoscopy, 218
Disk, blur, 190, 193
Disk, central diffraction, 193
Disk, Placido's, 39, 40
Dispersion, 5
    relative (nu value), 60
Distortion, 188
    barrel, 189
    of ophthalmic prisms, 74
    pincushion, 190
Duane line, 195

## E

Effective power, 113
    of the front surface, 145
Electromagnetic radiation, 19
Ellipse, Tscherning, 211; *See* Tscherning ellipse
Emmetropia, 102
Entrance-exit pupil centers, 192
Entrance pupil, 137, 138, 191
Exit pupil, 191
Extinction coefficient, 25
Eye movement and ophthalmic lenses, 198

## F

Far point, 102
    of hyperopic eye, 108
    of myopic eye, 104
    sphere, 209, 210
Field, curvature of, 187
    depth of, 194

Field of view decreased through a prism, 73
Filters, 23ff
    Corning glass, 23
    interference, 23
    neutral-tint, 23
    Wratten gelatin, 23
Flux, luminous, 240
f-number, 247
Focal length, 38, 40
    anterior, 151
    back, 142
    posterior, 150
Focal lines, 163
    radial (sagittal), 185
    tangential, 185
Focal plane, 86
Focal points, 85
Focus, depth of, 193
Foot-candle, 242
Foot-lambert, 244
Fraunhofer lines, 16, 183
Free stereoscopy, 232

## G

Galilean afocal lenses, 148, 197
Galilean telescopic system, 123
Gelatin filters (Wratten), 23
Geometric axis, 168
Geometric cylinders, 168
Geometric optics, 6
Grating, diffraction, 14
Gullstrand core lens, 156
Gullstrand schematic eye, 155

## H

Haploscope (Hering), 231
Hertz, 18
Hruby lens, 236
Hues, 5
Huygens, 4, 9
Hyperfocal distance, 194
Hyperopia, 107
Hyperopic eye, far point of, 108

# I

Illuminance, 242
Illumination, 239
Imagery, blurred, 190
Imagery, stigmatic, 29, 162
Images, optical, 27ff
  real, 29, 35
  virtual, 29, 34
Index of refraction, 60
  absolute, 51
  measurement of, 60
  relative, 51
Interference, 9
  filters, 23
Interval, Sturm's, 165
Inverse-square law, 243

# K

Knapp's rule, 205

# L

Lambert, 244
Lambert's law, 24
Law of sines, 129
Lens, 84ff
  afocal, 148
  afocal magnifying, 206
    meridional, 207
    overall, 207
  bitoric, 207
  contact, 214
  core, 156
  corrected-curve ophthalmic, 209
  cross-cylinder test, 176
  cylindric, components of, 171
    scissors effect of, 177
  diverging, 90
  for subnormal vision, 99
  Hruby, 236
  ideal converging, 88
  ideal thin, 88
  iseikonic, 148, 206
  magnifying, 97
  meniscus type, 143
  ophthalmic, with eye movement, 198
  power of, 88; *See* Power of a lens
    effective, 113
  position of, 109
  shape factor of, 139, 145
  thick, 102, 116, 127, 138
  Tillyer, 212
  trial-case test, 208
  vertex of, 143
Lensometer, 227
Light, 3
  dispersion of, 5
  monochromatic, 6
  pencil of, 27
  polarization of, 16
  reflection of, 22
  refraction of, 49
  scattered, 22
  transmission of, 23
  velocity of, 4
  wave theory of, 4, 7
    phase, 9
Light beam, 27
Light interference, 9
Light meter, 242
Lumen, 240
Luminance, 243
Luminance scale, 245
Luminous efficiency, 242
Luminous flux, 240
Lux, 242

# M

Maddox rod, 177
Magnification, 38, 42, 93, 96, 132
  angular, 62, 96
  chromatic difference in, 190
  from single refracting surfaces, 132
  of ophthalmic lenses used with the eye, 197
  of ophthalmoscope, 101
  power factor in, 199, 204
  shape factor in, 200, 204
Malus, law of, 18
Maxwell, 18
Meridian, base-apex, 65
Microlambert, 244
Millilambert, 244
Millimicron, 16

Minimal angle of resolution, 192
Minimal deviation, 67
Mirror, 30ff
  concave, 41
  convex, 36
  plane, 30, 32
Molecular absorption coefficient (extinction coefficient), 25
Myopia, 104
Myopic eye, far point of, 104

## N

Newton, 4, 5
Newton's fringes, 10
Newton's rings, 12
Nicol prism, 17
Nodal points, 149, 150, 192
  location of, 153
Nu (relative dispersion), 60

## O

Ocular refraction, 102
Oculometer, 232
Opacity, 23
Opaque object, 21
Ophthalmometer, 39
Ophthalmoscope, magnification of, 101
Ophthalmoscopy, 218
  direct, 218
  indirect, 219
Optical system, afocal, 122
Optical system, aplanatic, 29
Optical system, speed of, 255
Optics, geometric, 6
Optics, physical, 6
Optometers, 236

## P

Pencil of light, 27
Photometer, 255
Photometric units, 239
Photometry, 256
  heterochromatic, 256
Physical optics, 6
Pincushion distortion, 190
Pinhole, 195
Placido's disk, 39, 40
Planck, 19
Plane mirror, 30, 32
Plane parallels, 61, 62
Polarization of light, 16
Polaroid, 17
Pole of a surface, 37, 133
Power factor, 199, 204
Power of a lens, 88, 109
  effective, 113
    of the front surface, 145
  true, 120, 142
  vertex, 142, 143, 145, 198
Prentice, 69
Principal planes, 118, 138
Principal points, 118
  first, 119, 152
  second, 119, 152
  positions of, 140
Prism, 64ff
  decreased field through, 73
  deviating power of, 66, 67
  Nicol, 17
  ophthalmic, 64
    aberration of, 72
    distortion of, 74
  principal section of, 64
  rotary (Risley), 78
Prism diopter, 69
Prism powers, 69
Prismatic deviation, 66, 67, 69, 70, 199
  effect of tilt, 72
  measure of, 69
  minimal, 67
  resolution of, 75
Prismatic displacement, 201
Pupil, entrance, 137, 138
Pupil, entrance-exit centers of, 192
Pupil, exit, 191
Purkinjé reflexes, 46

## Q

Quantum theory, 19

## R

Radiant energy, 3
  spectral distribution of, 20
Radiation, electromagnetic, 19
Radiation, black-body, 19, 240
Rays, 6, 27
  chief, 87
  extra-axial, 87
  paraxial, 46, 87, 130
    tracing of, 136
  skew, 87
Reflectance, 29
Reflection of light, 29
  diffuse, 29
  law of, 30
  regular, 29
  specular, 29
Refraction, 49
  by spherical surfaces, 127
  index of, 60
    absolute, 51
    measurement of, 60
    relative, 51
  ocular, 102
Refraction of light, 49
Refractive error, 104
Resolution, minimal angle of, 192
Resolution of prismatic deviations, 75
Resolving power, 192
Retinal illuminance, 244
Retinal image, size of, 157, 202
Retinoscope, 220
Retinoscopy (skiametry), 220ff
Risley prism, 78
Römer, 4
Rotary deviation, 207

## S

Scheiner test, 234
Schematic eye, 155ff
  cardinal points of, 157
  dioptric constants of, 157
  Donders, 155
  Gullstrand, 155
  Listing, 155
  reduced, 158
  Stenström, 155
  von Helmholtz, 155
Shape factor of lens, 139, 145
Sign convention, 43
Sines, law of, 129
Skiametry, 220
Slit, stenopaic, 195
Snell's law, 49ff
Source intensity, 240
Spectral distribution of radiant energy, 20
Spectral lines, first-order and second-order, 14
Spectrum, continuous, 16
Spectrum, electromagnetic radiation, 18, 19
Spectrum, line, 16
Spectrum, visible, 16
Spectrum, visual, 5
Speed of the optical system, 247
Spherical equivalent, 165
Spherical surfaces, refraction by, 127
Stenopaic slit, 195
Stereoscope, 229
  Brewster, 230
  Wheatstone reflecting, 229
Stigmatic imagery, 29, 162
Stiles-Crawford phenomenon, 246
Stop, 188
Sturm's conoid, 166
Sturm's interval, 165
Sturm's lines, 163

## T

Telescopic system, 122
  astronomical, 123
  Galilean, 123
Terrascopic system, 125
Tillyer lens, 212
Toric surface, 162
Translucent objects, 21
Transparent objects, 21
Transposition, 170
Troland, 245
Tscherning ellipse, 210, 211

Ostwald branch, 211
Wollaston branch, 210

**V**

Vectographs, 17, 232
Vergence, 82
Vertex of the lens, 143
Vertex power, 142, 143, 145, 198
Visual acuity, 192

**W**

Wave theory of light, 4, 7
Wavelength, 8
Wheatstone reflecting stereoscope, 229
Wratten gelatin filters, 23

**Y**

Young, 9